The Art, Science, and Strategy of Longevity

AN EXPANSIVE EXPLORATION OF
AGING, HEALTH, AND HUMAN POTENTIAL

The Art, Science, and Strategy of Longevity

An Expansive Exploration of Aging, Health, and Human Potential

Ioulia Howard, MD, and
Don Howard, MD, PhD

Revised Edition (2026)

Copyright © 2025, 2026 by Ioulia Howard and Don Howard
All rights reserved.

No part of this publication may be reproduced, stored in a retrieval system, or transmitted in any form or by any means—electronic, mechanical, photocopying, recording, or otherwise—without prior written permission of the publisher, except for brief quotations in reviews or scholarly works.

First edition published in 2025.
This revised edition published in 2026.

Published in the United States of America by
Vibrant Ages Publishing
Mercer Island, Washington

ISBN: 979-8-9942516-2-1 (Hardcover)
ISBN: 979-8-9942516-1-4 (Paperback)
ISBN: 979-8-9942516-0-7 (eBook)

Library of Congress Control Number: 2025927569

Additional information and related resources are available through Vibrant Ages Publishing. For inquiries, contact: time@vibrantages.com

About This Revised Edition
This revised edition incorporates editorial refinements to improve clarity, readability, and distribution compliance. The book's structure, themes, and perspective remain unchanged. Select passages have been reframed to reduce prescriptive clinical detail while preserving the substance and intent of the original work.

Note to the Reader
Written for a general audience, this book explores aging, health, and longevity from scientific, clinical, and philosophical perspectives. It is intended for educational and informational purposes and does not provide medical advice, diagnosis, or treatment recommendations, nor does it create a physician—patient relationship.

The material reflects current thinking on aging and longevity and emphasizes understanding and interpretation rather than instruction. Scientific knowledge continues to evolve, and individual circumstances vary. Decisions related to health and medical care should therefore be made in consultation with qualified healthcare professionals familiar with one's personal situation.

This book may reference medications, supplements, diagnostic approaches, and interventions, including uses that are investigational or off-label. Nothing in this book should be interpreted as a claim that any practice, strategy, technology, or intervention can prevent, treat, reverse, or cure aging or disease.

*"Do not grow old, no matter how long you live.
Never cease to stand like curious children before the great
mystery into which we were born."*
—Albert Einstein

Preface

Aging touches every life, stirring curiosity and uncertainty in equal measure. Long perceived as inevitable decline, growing older is increasingly understood as a dynamic process influenced by genetics, environment, and the choices made over time. We stand today at a crossroads, where advances in medicine, biotechnology, and longevity science are transforming what it means to age. As these developments reveal new possibilities, they also raise new questions, underscoring the importance of guidance in a world where scientific progress often outpaces public understanding.

Healthy aging extends beyond biological markers; it encompasses emotional, social, and psychological dimensions as well. Across years of clinical experience, we have seen how neglecting these factors can affect well-being in interconnected ways. We have also come to recognize something equally consistent: when aging is approached deliberately—through psychological resilience, sustained social connection, and purposeful engagement—people often retain a striking sense of perspective, continuity, and fulfillment.

These clinical insights resonate with what we have observed in our own families. Between us, we have known multiple generations—five for Ioulia and seven for Don—spanning great-grandparents to great-grandchildren. These relationships have reinforced a simple truth: growing older is not just about adding years, but about defining the quality and depth of the years lived. Over time, a familiar pattern emerges: we do not grow old all at once; we grow old through the choices we repeat, and one day we recognize the life those choices have created.

Aging is neither a script we must follow nor a condition to escape. It is an experience shaped gradually through daily decisions, personal circumstances, and the knowledge accumulated along the way. What matters is not only living longer, but living with clarity, adaptability, and meaning, remaining actively engaged with the years ahead.

Inspired by decades of clinical practice, scientific inquiry, and reflection, this book grew out of experiences rooted in both our professional and personal lives. By weaving together insights from science, medicine, and philosophy, we offer this work as a steady companion for thoughtful exploration of aging and longevity.

We once viewed aging primarily as fate. Now, we see it as possibility.

Time is not inevitably lost—it is shaped through intention, moment by moment.

Through these moments, we craft the story of our lives.

With gratitude, we invite readers to explore the journey ahead.

—Drs. Ioulia and Don Howard

Table of Contents

Preface vii

Acknowledgments xiii

Introduction xv

Part I: Understanding Aging—History, Philosophy, and Science 1

Chapter 1   Aging Across Time: A Historical Perspective 3

Chapter 2   The Art of Aging Well: Philosophy, Time, and Meaning .. 17

Chapter 3   Why We Age: Scientific Theories on Life's Fundamental Mystery 27

Part II: Foundations of Longevity—Core Strategies for Health and Vitality 47

Chapter 4   Movement and Longevity: Motion Across a Lifetime 49

Chapter 5   Nutrition for Longevity: The Science Behind Food and Health 65

Chapter 6   Supplements and Functional Foods: Precision Nutrition and Aging 73

Chapter 7   Fasting and Caloric Restriction: What Happens Inside the Body 85

Chapter 8   Sleep and Aging: More Than Rest 95

Chapter 9   Stress, Resilience, and the Aging Mind 105

Chapter 10	Emotional and Spiritual Dimensions of Aging	111
Chapter 11	The Pleasure Principle: Sexual Intimacy and Aging	115
Chapter 12	Hormones and Aging: Timing, Balance, and Biological Change	123
Chapter 13	Hormone Replacement Therapy: Evidence, Risks, and Judgment	133

Part III: Rewriting the Rules of Aging—Longevity Science and Biohacking 147

Chapter 14	Rapamycin and Rapalogs: Inside a New Era of Longevity Research	149
Chapter 15	Metformin: Reimagining a Diabetes Drug for Longevity	161
Chapter 16	NAD$^+$ and Aging: NMN, NR, and Cellular Energy	171
Chapter 17	Targeting Aging at the Cellular Level: Senolytics, Metabolism, and Telomeres	177
Chapter 18	Non-Pharmacological Biohacking: Tools and Approaches Beyond Medication	185

Part IV: Next-Generation Longevity—Personalized Health and Revolutionary Innovation 199

Chapter 19	Regenerative Medicine: Emerging Pathways in Human Repair	201
Chapter 20	Gene Therapy and Aging: Transformative Ideas in Genetic Science	211
Chapter 21	Technology and Longevity: Tools Changing How We Age	219

Chapter 22	Artificial Intelligence and Aging: Promise, Precision, and Limits	231
Chapter 23	Partnering with Your Healthcare Provider: Building a Longevity Team	243
Chapter 24	Longevity by Design: Building a Personal Strategy	247

Postscript: The Age of Possibility: A Final Reflection 255

References 257

Acknowledgments

We are deeply grateful to Ioulia's parents, Alexander and Antonina, for their unwavering love, support, and encouragement throughout this journey. The time we spent with them in Russia—particularly during the winter months—offered a rare period of focus that allowed us to immerse ourselves in research and devote ourselves fully to writing this book.

We also thank Melissa Butterworth, whose thoughtful suggestion sparked the idea for this project and helped set it in motion.

Introduction

"Aging is an extraordinary process where you become the person you always should have been."
—David Bowie

Aging is humanity's oldest enigma—a puzzle that cultures have pondered, scientists have studied, and philosophers have contemplated for millennia. It is both universal and deeply personal, defining entire societies and individual destinies. Aging continues to inspire fascination and provoke fundamental questions about what it means to grow older. Why do we age, and what does aging actually entail? How have our perceptions of aging changed over time, and how might we engage more deliberately with this universal human experience at a moment when scientific discovery is advancing at an unprecedented pace?

Historically, societies have grappled with aging in diverse ways. Some cultures revered elders as keepers of memory and wisdom; others viewed aging primarily through vulnerability and decline. Philosophers—from antiquity onward—have examined the tension between accumulating understanding and undergoing physical change, offering enduring insights into aging's meaning and limits.

Today, thinking about aging is being reshaped by advances in science and medicine. Decades of research have deepened our understanding of the biological mechanisms associated with aging, clarifying cellular processes, genetic influences, and interactions among environmental and behavioral factors. Aging is increasingly described not as inevitable wear and tear, but as a multifaceted process arising from multiple interacting influences. This broader view changes how aging is understood.

Later life is also a human narrative, formed not only by biology but by experience, reflection, and the shifting contours of a life over time. This book follows that story across multiple dimensions, moving beyond the question of longevity toward an understanding of how resilience, clarity, and purpose develop. It begins with the historical, philosophical, and biological foundations that define aging, then turns to the practical foundations that support daily life—movement, nutrition, sleep, stress resilience, emotional well-being, and hormonal balance. The exploration widens further to emerging scientific frontiers—biohacking, regenerative medicine, and new technologies—that expand what aging may encompass. Because these areas intersect, key ideas recur across chapters as part of an evolving dialogue, each return adding nuance and reflecting the dynamic nature of longevity science.

Aging's significance extends beyond historical reflection or scientific inquiry; it emerges through the intentions, values, and aspirations that guide a life. This book is neither a purely academic treatise nor a clinical handbook. Rather, it offers a deeper way to understand aging and to consider one's own path through time. The stakes are high, the possibilities extraordinary—and the choices made in the present ripple outward, influencing the character of the years ahead and the world carried forward by the lives that follow.

Part I
Understanding Aging—History, Philosophy, and Science

CHAPTER 1

Aging Across Time: A Historical Perspective

"To know the past is to understand the present."
—Pearl S. Buck

Aging is one of humanity's deepest truths, defining every life and influencing every society's understanding of time. Throughout history and across cultures, our attitudes toward aging have swung dramatically—oscillating between reverence and dread, celebration and resistance. Some societies honor their elders as treasured living libraries, vessels of wisdom whose insights illuminate pathways for younger generations, preserving collective memory and identity. Others perceive aging predominantly in terms of loss—a gradual fading of strength, beauty, and relevance—sparking an enduring quest to delay, mitigate, or even defy its effects.

These contrasting perspectives have defined cultural ideals, traditional norms, and medical progress. Civilizations uncomfortable with aging's inevitability have tirelessly sought remedies that promise renewed vitality, from ancient herbal elixirs to today's advanced biotechnology aimed at understanding the cellular processes associated with aging. Amidst this ceaseless pursuit of longevity, a question emerges: as our scientific understanding of aging continues to grow, perhaps our most compelling challenge lies not in lengthening life's narrative, but in deepening the wisdom and wonder within each chapter.

Today, as rapid scientific discovery broadens our understanding of aging in ways once unimaginable, revisiting historical attitudes toward growing older takes on renewed importance. Reflecting on how earlier cultures grappled with this universal experience offers essential context for contemporary explorations of what aging well can signify. Seen across time, aging appears not as fixed biological fate, but as an ever-unfolding cultural story—continually influenced by human imagination, enduring values, and an abiding curiosity about life's possibilities.

The Earliest Views on Aging

In humanity's earliest societies, aging carried an evocative duality—deeply revered for its wisdom, yet shadowed by undeniable fragility. Elders were the foundation upon which their communities stood, valued not only as survivors but as essential keepers of lived experience. They held intimate knowledge of nature's rhythms, anticipating shifting seasons, identifying life-sustaining plants, healing sickness, and sensing hidden dangers long before they emerged. Through shared stories and carefully preserved insights, elders wove threads of continuity, guiding their communities from one generation to the next. Even as they illuminated the path forward, their growing vulnerability served as an ever-present reminder of life's fleeting beauty and time's steady, relentless progression.

As civilizations flourished, these early reflections on aging deepened and found expression in myths, poetry, and epic narratives. Few stories capture humanity's eternal tension with mortality as poignantly as *The Epic of Gilgamesh*. Within this ancient Mesopotamian masterpiece, Gilgamesh—a figure mighty yet touchingly human—is devastated by the loss of his cherished companion, Enkidu. Grief-stricken and haunted by the certainty of his own mortality, Gilgamesh embarks on a desperate and arduous quest for immortality.

The eternal life he longs for remains just beyond his reach. Instead, Gilgamesh uncovers a truth as timeless as his story: physical immortality is an illusion. True endurance arises not from conquering aging, but from the legacies we cultivate—the wisdom we generously share, the communities we lovingly nurture, and the stories we purposefully pass along. Gilgamesh's transformative journey continues to resonate deeply, reaffirming humanity's oldest wisdom: aging and death are not foes to conquer, but essential currents in the ever-flowing river of human existence.

Gilgamesh's story invites us to reflect on our own lives. As time forms us, what legacy will we choose to leave behind? How will the years we've been given transcend the limits of our own existence?

Aging in the Ancient World: Wisdom, Decline, and Social Realities

In the ancient world, aging represented a profound duality—revered for its wisdom yet shadowed by inevitable physical decline. Venerated philosophers such as Socrates, Plato, and Aristotle saw advanced age as far more than biological fate; they viewed it as an extraordinary opportunity for intellectual growth and ethical reflection. Aristotle, in particular, considered later life uniquely suited for the cultivation of reason and discernment—faculties polished and sharpened by a lifetime of experience. Even so, he openly acknowledged a bittersweet trade-off: with wisdom's bounty came the certain toll of bodily deterioration.

This philosophical recognition of aging's dual nature—wisdom paired inseparably with decline—was vividly embodied in Greek mythology and symbolism. Central to Greek cultural imagination stood Chronos, the personification of relentless, unstoppable time. His presence was a compelling reminder of human mortality, echoed powerfully through myths and legends. At the same time, these same myths expressed humanity's timeless yearning for renewal and immortality—seen in tales of rejuvenating fountains and restorative elixirs. Such stories reveal the ancient conflict between accepting aging's inevitability and the enduring human dream of escaping its boundaries, a tension that remains palpable today in both medical innovation and popular culture.

In ancient Rome, this dynamic played out differently, influenced by class and social privilege. Among the Roman elite, longevity was interpreted as evidence of divine favor or moral achievement—a tangible reward for virtuous living. The distinguished philosopher Cicero eloquently captured this ideal in his seminal work, *De Senectute* (*On Old Age*). To Cicero, aging was not a slow decline into irrelevance but rather an exceptional stage of life, a time for refining character, deepening wisdom, and securing a meaningful legacy. He viewed old age as dignified and enriching—provided one approached it intentionally, with virtue and continued intellectual engagement.

Beneath Rome's philosophical ideals lay harsher realities for those without privilege: the poor, enslaved, and marginalized. Lacking wealth, status, and education, older members of these communities typically endured intensified hardship, vulnerability, and social exclusion. For them, aging was rarely a serene period of philosophical reflection or community reverence; instead, it was marked by frailty, economic uncertainty, and dependence. Their struggles underscored stark inequalities deeply embedded in ancient societies, highlighting how strongly social status shaped individual experiences of growing older.

This enduring tension—aging as both a gateway to wisdom and an inevitable source of decline—continues to resonate in the present. Ancient philosophers illuminated the aspirational dimensions of growing older, even as their societies revealed the harsher truths of aging. Today, this historical perspective challenges us anew, prompting an essential question: How can we ensure dignity, wisdom, and vitality define aging—not only for the privileged few, but for everyone?

Aging in the Medieval World: Wisdom, Faith, and Social Realities

In the medieval world, aging meant far more than a biological certainty—it was intricately interlaced with spiritual beliefs, cultural traditions, and the rhythms of daily life. Across diverse medieval cultures, old age represented not decline, but transformation: a sacred opportunity for introspection, spiritual maturity, and preparation for what lay beyond. Within Hindu and Buddhist traditions, aging signified an intentional shift from worldly attachments toward spiritual awakening. Central to this view was *samsara*, the eternal cycle of birth, death, and rebirth. Physical decline was not viewed as misfortune but rather as a poignant reminder of life's ephemeral nature, encouraging detachment from material desires. Ascetics, monks, and sages who purposefully withdrew from society in pursuit of deeper truths were revered as guiding lights, embodying wisdom, discipline, and spiritual liberation.

Christianity, too, imbued aging with deep spiritual meaning, though in a distinct way. For medieval Christians, later years became a sacred stage—a period dedicated to strengthening one's bond with God, reflecting on life's moral choices, and readying the soul for eternity. Biblical figures such as Moses, who guided his people with strength despite advanced years, and Simeon and Anna, the wise elders who recognized the infant Christ as the long-awaited redeemer, exemplify the deep grace, spiritual discernment, and respect attributed to old age. Nevertheless, Christianity stopped short of idealizing aging entirely. Instead, it openly acknowledged physical frailty and suffering as humbling reminders of life's transitory nature. Consequently, bodily decline was reframed as a sacred journey, purifying and readying the soul for eternal redemption.

Aging was understood through more than just the cultural filter of religion; it was also experienced amid complex social realities. Elders occupied indispensable roles as guardians of collective wisdom, maintaining stability, transmitting knowledge, and safeguarding traditions. Especially in rural communities where formal education was rare, older adults offered mentorship—teaching younger generations crucial skills for survival, farming, healing, and preserving history through oral traditions. Their steady presence provided continuity and reassurance in times marked by war, famine, and upheaval.

Despite these respected roles, longevity remained exceptional—and consequently, somewhat precarious. Medieval life, often harsh and uncertain, was marked by disease, violence, and scarcity, resulting in relatively short lifespans. Those who reached advanced age thus occupied an ambiguous position within society. Longevity might reflect divine favor, moral virtue, or exceptional resilience, inspiring admiration and reverence. However, particularly during societal crises, older adults—often women—could provoke suspicion, distrust, or even accusations of witchcraft. Their very existence challenged prevailing norms, revealing deep-seated anxieties surrounding aging's enigmatic nature.

This inherent duality—aging as both revered wisdom and troubling vulnerability—captures the complexity of medieval conceptions. Aging during this era was neither purely biological nor exclusively spiritual; rather, it unfolded at the crossroads of religion, culture, economics, and personal circumstance. Reflecting upon these historical insights deepens our contemporary understanding, reminding us that modern attitudes toward aging remain forged by centuries of evolving beliefs and social conditions. Thus, exploring the medieval experience of aging reveals not a narrative of either spiritual insight or physical decline, but a layered portrait of elders navigating reverence, uncertainty, and adversity. Their lives, etched by perseverance and adaptability, illuminate the timeless human capacity to create meaning even amid intense hardship and uncertainty.

The Renaissance and Enlightenment: Aging in a New Light

The Renaissance transformed society's view of aging, inspired by a renewed fascination with humanism and the complexities of lived experience. Across Europe, as art blossomed, science advanced, and philosophy deepened, aging emerged prominently as a subject of artistic admiration and intellectual reflection. Master artists, notably Leonardo da Vinci and Michelangelo, captured the aging body with exquisite sensitivity—faces etched by wisdom, hands dignified by labor, forms softened by years. Rather than depicting decline, their work celebrated aging as durability, dignity, and depth. In elevating old age from biological inevitability to aesthetic reverence, these visionaries endowed it with enduring philosophical significance.

This artistic awakening ran parallel to groundbreaking scientific inquiry, marking a pivotal shift in understanding human aging. The body, once considered a sacred and immutable mystery, became an object of careful, systematic investigation. Anatomists such as Andreas Vesalius ventured into previously uncharted territories through detailed dissections, laying the foundations of modern physiology. While their grasp of aging's intricate biology remained limited, their work signified a critical conceptual shift: aging was no longer viewed solely as fate—it had become a phenomenon open to exploration, observation, and scientific inquiry.

The Enlightenment deepened and broadened this intellectual transformation, championing reason, human potential, and lifelong intellectual growth. Philosophers like Voltaire and Rousseau challenged prevailing views that equated aging with decline. Instead, they reframed later life as uniquely suited for reflection, intellectual discovery, and personal growth. Rousseau envisioned old age as liberating—a phase where one could step away from the distractions of youth to contemplate deeper philosophical truths. Similarly, Voltaire remarked, "The longer we dwell on our misfortunes, the greater is their power to harm us," suggesting that aging, approached thoughtfully, could become a time of genuine personal fulfillment. Enlightenment ideals thus nurtured a powerful belief: intellectual vitality could not only persist but flourish well into the later chapters of life.

Underlying this optimism lay emerging cultural tensions, tensions that continue to shape contemporary attitudes. Even as elders retained their respected status as guardians of wisdom, Enlightenment values emphasizing innovation and relentless progress began to associate vigor and productivity more closely with youth. The rise of industrialization further intensified this shift, prioritizing attributes such as physical strength, adaptability, and efficiency—qualities often attributed primarily to the young. Gradually, older adults came to be viewed not just as repositories of valuable historical memory and life experience but also as symbols of stagnation, inefficiency, and obsolescence. Thus, industrialization quietly planted the seeds of modern ageism, establishing a cultural bias favoring youth that persists to this day.

The Renaissance and Enlightenment stand as transformative eras in humanity's ongoing dialogue about aging. Old age was simultaneously celebrated for wisdom and potential intellectual fulfillment, yet increasingly scrutinized as a hindrance to innovation and progress. This duality, deeply embedded within contemporary society, continually challenges us to reconcile reverence for age-earned wisdom with the societal idealization of youthfulness. How we navigate this tension—whether by reinforcing unspoken ageist biases or embracing aging as an enriched, meaningful stage of life—remains a complex yet crucial question. It is a conversation that began long ago, crafted by the voices of artists, philosophers, and scientists whose insights still resonate today.

The Industrial Revolution: Aging in a Changing World

The Industrial Revolution transformed society, altering the meaning and experience of aging. As agrarian lifestyles yielded to industrial economies, traditional family structures—once the cornerstone of elder care and support—began to unravel. Historically, rural elders had served as custodians of cultural wisdom, mentors guiding younger generations, and stabilizing forces within their communities. As younger family members flocked to rapidly expanding cities in pursuit of employment, many older adults found themselves left behind, their roles diminished and their societal value eroded. Even those elders who followed faced unanticipated struggles, navigating urban landscapes devoid of the familiar intergenerational bonds that had long sustained them. In this new social reality, the fabric that had safeguarded older generations began to fray, leaving many increasingly vulnerable to isolation and neglect.

At the heart of industrialization was an uncompromising drive toward efficiency, productivity, and relentless economic growth. Within this shifting landscape, aging—inevitably accompanied by gradual physical and cognitive changes—began to lose its traditional association with wisdom and respected experience. Older workers, struggling to keep pace with the harsh demands of factory labor, often found themselves marginalized and perceived as inefficient or expendable. This shift was more than an economic adjustment; it marked a cultural transformation, recasting aging from a respected, natural stage of life into an apparent obstacle to progress. Here were planted the seeds of modern age-based bias—attitudes favoring youth and productivity that still echo through societal perceptions today.

The legacy of the Industrial Revolution was not only one of marginalization. The same era that disrupted traditional support networks also ignited an unprecedented exploration of aging. The nineteenth century brought extraordinary advances in medical knowledge and saw the emergence of gerontology as a systematic field of study. Researchers moved beyond philosophical speculation, diving deeper into concrete biological processes—cellular degeneration, declining organ functions, and factors influencing longevity. These pioneering studies aimed not only to better understand lifespan but to examine the quality of life experienced throughout aging, laying foundations for modern longevity science. Concurrently, widespread public health initiatives dramatically reduced infectious diseases that disproportionately impacted older populations, fundamentally altering the lived experience of aging.

The Industrial Revolution left behind a complex, dual legacy. On one hand, it dismantled established intergenerational bonds, framing aging as a hindrance to economic efficiency. On the other, it spurred groundbreaking scientific progress, recasting the concept of aging from inevitable decline to a dynamic, potentially modifiable phenomenon. This dialectic—between marginalization and innovation—still influences contemporary discourse around aging. Today, society remains poised between reverence for elder wisdom and deeply entrenched biases favoring youthfulness. This nuanced debate, with its distinctly modern challenges, can trace its roots directly to the transformative upheavals of the Industrial Revolution.

IOULIA HOWARD AND DON HOWARD

The 20th Century: Aging as a Scientific Frontier

In the 20th century, aging evolved from life's inevitable conclusion into an expansive landscape of scientific discovery and potential. Breakthroughs in medicine and public health—antibiotics, vaccines, improved sanitation, and better nutrition—sparked unprecedented gains in human lifespan. Aging ceased to be viewed as a fixed biological destiny and instead emerged as a promising realm ripe for investigation, deeper understanding, and possible transformative intervention.

Scientists began unraveling the cellular mysteries of aging, identifying processes such as oxidative stress—the damage inflicted by unstable, oxygen-rich molecules—as well as DNA deterioration and telomere shortening, the gradual erosion of protective chromosome end caps. Genetic research further illuminated these complexities, revealing longevity-associated genes like FOXO, which bolster cellular resilience, and metabolic pathways such as mTOR, a critical regulator of cell growth and metabolism. For the first time in history, a provocative question took center stage: could aging be slowed or even reversed? From this daring inquiry arose the dynamic field of bio-gerontology, committed not just to extending lifespan, but to enriching healthspan—the years spent largely unburdened by chronic disease or disability.

Even as science expanded the boundaries of longevity, societal perceptions of aging grew increasingly complex—and often contradictory. Older adults were celebrated for their extraordinary contributions to history, politics, and culture, exemplified by towering figures such as Winston Churchill, Eleanor Roosevelt, and Pablo Picasso. These individuals embodied the inspiring idea that creativity, leadership, and influence could flourish well beyond youth, challenging traditional notions of age-related decline. Simultaneously, however, cultural trends fueled by media increasingly idealized youth, positioning it as the ultimate standard of beauty, vitality, and productivity. Industries like fashion, advertising, and entertainment reinforced these ageist stereotypes, marginalizing older adults, particularly in workplaces where seasoned experience frequently yielded to preferences for youthful vigor.

As global populations grew older, societal structures evolved in response, prompting new policies and support systems designed for an aging world. Initiatives such as the establishment of Social Security in the United States, the rise of retirement communities, and the proliferation of senior-care facilities sought to provide financial security, accessible healthcare, and social engagement. However, these developments also laid bare persistent social tensions, prompting debates about dignity, autonomy, and equity—issues that resonate deeply in contemporary conversations about aging.

The 20th century's legacy is a paradoxical blend of scientific triumph and cultural ambivalence. Medical advancements ignited optimism that both life and vitality could be prolonged, even as shifting societal values exposed contradictions—simultaneously honoring the wisdom of older adults while idolizing youth. This transformative era positioned aging as an intricate intersection of scientific discovery and cultural dialogue, compelling contemporary society to grapple with the ethical, social, and medical complexities embedded in longevity.

The 21st Century: The Transformation of Aging

Today, aging inhabits a remarkable moment of evolution, defined by extraordinary leaps in science, technology, and shifting cultural perceptions. Once considered life's inescapable script, aging has become humanity's most intriguing narrative—dynamic, adaptable, and open to reinterpretation.

Scientific advances have illuminated the complex molecular dance underlying the aging process, revealing mechanisms such as cellular senescence, mitochondrial change, and epigenetic drift. These insights have helped shape the modern fields of geroscience and regenerative medicine, which investigate the fundamental biology of aging. Breakthrough technologies—from precision gene editing (CRISPR) and AI-driven diagnostics to stem-cell—based research and senolytic investigations—aim to deepen our understanding of how biological systems evolve over time.

Rather than offering definitive promises about lifespan or health outcomes, these emerging approaches open new avenues for studying the processes that influence aging. The innovations defining this landscape, and the science guiding them, will be explored in detail in the chapters ahead.

Parallel to these medical innovations, society's perception of later life is undergoing a transformation. Contemporary culture increasingly celebrates older adulthood as a flourishing chapter of continued creativity, reinvention, and meaningful contribution. Movements emphasizing "successful aging" encourage older adults to defy traditional stereotypes—launching entrepreneurial ventures, crafting powerful art, and setting new athletic milestones. In doing so, they reimagine human potential, challenging conventional expectations of what life's later stages can hold.

Even amid these promising shifts, ageist biases linger stubbornly. Experienced professionals still encounter discrimination in workplaces that prioritize youth over wisdom. Media and entertainment frequently perpetuate simplistic portrayals of older adults, reinforcing perceptions of diminished capability and relevance. Within healthcare settings, assumptions about inevitable frailty often limit proactive care and prevent opportunities for improved quality of life. Such biases erode individual dignity and rob society of the insights, skills, and talents older generations have to offer.

Beyond these challenges, however, the 21st century has also ushered in greater recognition of the emotional and social dimensions of aging. Concepts like lifelong learning and intergenerational collaboration are dissolving artificial divides between age groups, nurturing mutual respect, empathy, and collective growth. Globally, communities are embracing age-inclusive initiatives—accessible housing, innovative smart-home technologies, supportive social programs, and intentionally designed public spaces—that empower older adults, preserving their independence, fostering engagement, and strengthening social connections.

Growing older today unfolds at the intersection of scientific discovery and cultural change, where long-standing views of later life meet new and rapidly advancing possibilities. Later adulthood is increasingly regarded as a period distinguished by reflection, accumulated experience, and the expanding influence of social and technological developments. In historical perspective, aging in the 21st century marks a turning point—where humanity's ongoing confrontation with mortality encounters profound scientific and cultural shifts, altering both the experience and interpretation of longevity.

The Next Chapter: Reimagining Aging for the Future

Aging has always been inextricably linked with the human story—informing cultures, traditions, and identities across countless generations. From ancient societies that honored elders as guardians of wisdom, to modern pursuits aimed at extending the human lifespan, our relationship with aging has continually shifted between reverence and resistance. Throughout history, we have embraced growing older as part of nature's cycle, confronted it as an obstacle to overcome, and pondered it as one of life's enduring mysteries.

Today, advances in biotechnology, medicine, and longevity science are unveiling groundbreaking discoveries that reconceptualize what it means to age—not as inevitable decline, but as a dynamic biological journey we can study, influence, and perhaps even transform. Alongside the potential for longer lives emerges a deeper, more challenging question: How do we ensure those years are defined not by survival alone, but by meaningful engagement, genuine autonomy, and a sense of purpose? In the end, longevity itself is not the ultimate measure. What matters is the depth, intention, and significance with which we embrace the time we have.

This shared journey extends beyond the physical realm—it helps form our identities, relationships, and the legacies we leave behind. It also defines how we perceive ourselves and how we connect with others, guiding the impressions we make and the stories we tell. At the heart of this passage through life lies a continual testing of character and an invitation toward adaptability, offering moments of introspection, renewal, and rediscovery.

In the end, we are prompted toward deep reflection: What does it mean to live with significance—not only in the vigor of youth, but across every stage of life?

As science deepens our understanding of aging, we increasingly appreciate that this universal journey reveals far more than biology—it illuminates our collective human narrative. History shows us how societies across time have grappled with the meaning of growing older, each generation adding new layers of interpretation and understanding. Aging becomes a story of continuity and reinvention, rooted in cultural values, societal structures, and evolving perceptions of life's meaning. Looking forward, the future of longevity draws on these historical insights, inviting us not simply to extend life but to deepen its substance—honoring the lessons of the past, engaging the possibilities of the present, and approaching tomorrow with curiosity, discernment, and renewed intention.

CHAPTER 2

The Art of Aging Well: Philosophy, Time, and Meaning

> *"The mind is its own place, and in itself can make a heaven of hell, a hell of heaven."*
> —JOHN MILTON, *PARADISE LOST*

Aging is more than a biological phenomenon; it is a uniquely human experience that unfolds across time and memory. It echoes through the corridors of our existence—profound, poignant, and endlessly thought-provoking—prompting us to consider some of the deepest mysteries of the universe: the fluidity of identity, the pursuit of purpose, and the inevitability of mortality. As science tirelessly strives to prolong life and safeguard health, philosophy urges a deeper contemplation: What does it mean to live well and fully throughout the entirety of our years?

Throughout centuries, great thinkers have grappled with this perennial inquiry, painting aging either as an enlightening pathway toward wisdom or as a gradual descent into decline. Some have envisioned growing older as a portal to richer understanding, a season when life's accumulated experiences distill clarity and strengthen character. Others, however, have depicted it as a quiet unraveling, an erosion of youthful vigor and social significance. How can such sharply contrasting perceptions coexist? What does this duality reveal about our ongoing effort to understand aging?

Philosophical thought has deeply influenced cultural attitudes toward aging, offering lessons about resilience, acceptance, and humanity's quest for purpose amidst change. Across diverse traditions, philosophy offers guidance not only in navigating life's later chapters but in embracing their transformative potential. Even amidst today's unprecedented medical and technological breakthroughs, these ancient teachings remain strikingly relevant, reminding us that longevity does not guarantee satisfaction or happiness.

If granted additional time, how would we spend it most meaningfully? What defines an existence not only lived but richly experienced? This chapter invites readers to explore philosophical traditions—drawing upon ancient wisdom and contemporary perspectives—to reimagine aging. Through these perspectives, aging emerges not only as a challenge to overcome but as an essential space for reflection, offering continual opportunities to clarify life's purpose and deepen our understanding of self.

Ancient Perceptions: How the Great Thinkers Saw Aging

The experience of growing older traces its arc through every human life, not as decline, but as a reckoning with time, identity, and the shifting contours of the self. Long before science sought ways to delay or reverse aging, philosophers grappled with its implications. Across cultures and centuries, a common understanding emerges: the later stages of life are not the gradual erosion of time but constitute a chapter rich with possibility, reflection, and transformation.

The Stoics approached aging with calm acceptance, viewing it with the same clarity and composure they brought to all of life's uncertainties. They recognized that advancing years are inevitable—utterly beyond our control—and thus unworthy of lament. Resisting this natural process, they believed, meant waging a futile battle against nature, inevitably leading only to frustration. Instead, welcoming maturity with grace fostered resilience, wisdom, and inner peace. Central to this outlook was the Stoic ideal of *amor fati*, or "love of fate," encouraging a wholehearted acceptance of life's later chapters rather than fear or avoidance.

Marcus Aurelius, reflecting in his *Meditations*, reminded himself daily of life's fleeting nature. By cultivating virtue and integrity rather than desperately clinging to youth, he sought to approach old age not with regret but with strength. Seneca went further still, boldly asserting that life is measured not by length but by depth—a timeless reminder that resonates perhaps more powerfully today than ever. For the Stoics, virtue—the highest good—meant living harmoniously with reason and nature, with the passage of years understood as integral to this harmony.

Where the Stoics saw life's progression as a test of resilience and virtue, Aristotle viewed it as life's culminating chapter—a fulfillment rather than a loss. His philosophy of *eudaimonia*, often translated as human flourishing, suggested that genuine meaning arises not from fleeting pleasures or youthful strength but from purposeful, reflective, and virtuous living. Aristotle understood virtue not as moral correctness but as a lifelong commitment to intellectual growth, ethical refinement, and thoughtful moderation. His Doctrine of the Mean emphasized balance: the careful path between excess and deficiency. Older adulthood, he believed, provides the ideal setting to achieve this equilibrium, tempering youthful impulsivity with seasoned insight.

Rather than a period of deterioration, Aristotle celebrated the later years as a stage when wisdom and character reach their fullest bloom. The twilight of life, he argued, offers space for deeper contemplation, richer relationships, and the fine-tuning of ethical pursuits. In striking contrast to modern culture's fixation on youth as the pinnacle of worth, Aristotle acknowledged the quiet grace that comes with experience—a subtler, more nuanced understanding of life's complexities.

While Aristotle highlighted intellectual cultivation, Eastern traditions introduced another dimension: harmony with nature rather than mastery over it. Taoism, through Laozi's teachings, portrays the process of growing older not as a problem needing solutions but as a natural unfolding to approach with ease. Central to this approach is *wu wei*—effortless action—encouraging alignment with life's flow rather than resistance. Like water gliding effortlessly around stones in a stream, Taoism suggests maturing gracefully means adapting smoothly and naturally, without conflict or force.

Buddhism, too, presents a vision rooted in acceptance. The principle of *anicca*, or impermanence, underscores the transient nature of all things. Clinging tightly to youth, vitality, or identity only magnifies suffering. Unlike Stoicism's disciplined acceptance, Buddhism views letting go as liberation—freedom from the pain of attachment. Buddhist teachings foster detachment, discernment, and serenity, reframing older age not as loss, but as an opening for compassion, understanding, and inner peace. Zen practitioners, through seated meditation (*zazen*), train the mind to release past regrets and future anxieties, fully inhabiting the present moment with calm clarity.

While Western traditions often emphasize autonomy, Confucian philosophy regards elderhood as a transition into respected leadership. Elders become guides and mentors, responsible for nurturing younger generations' moral and intellectual development. Central to this is *xiao*, filial piety, an ethical principle underscoring respect for elders as repositories of invaluable knowledge. Unlike modern individualistic societies, where advancing age is sometimes equated with diminishing relevance, Confucian culture cherishes older generations as living links between past traditions and future growth.

Zen Buddhism offers another shade of understanding—a more introspective exploration of life's later chapters. Rather than focusing on societal roles, Zen invites an intimate awareness of one's relationship with time. Its teachings promote radical presence, illuminating how suffering arises from attachments to the past or anxieties about the future. Zen philosophy encourages accepting each moment exactly as it unfolds, allowing the natural progression of life to occur effortlessly. Mindfulness and meditation thus illuminate the later years not as something feared but as an open invitation toward clarity, acceptance, and equanimity.

These ancient perspectives challenge today's pervasive fear of growing older, offering timeless lessons about fortitude, purpose, and serenity. They encourage us to reconsider this stage of life not as deterioration but as continuing evolution—an ongoing process of growth, adaptation, and refinement. Taken together, these philosophies portray aging not as a slow retreat from life but as an artful collaboration with time—encouraging us to harmonize gracefully with change and draw meaning from the wisdom of experience.

The Modern Age: Aging in an Era of Scientific Progress

For most of human history, aging was an unyielding reality—fixed firmly by fate, fortune, and nature's immutable order. Today, that certainty feels less absolute. Scientific advances once relegated to the realm of speculation now offer new ways of studying aging at its most fundamental levels. Regenerative approaches explore how tissues sustain themselves over time; gene-focused technologies provide tools for examining cellular pathways; pharmaceutical research investigates mechanisms associated with senescent cells; and molecular studies illuminate how the body maintains its own systems of repair. The landscape is striking—promising not answers, but possibilities, and expanding the ways we think about longevity, vitality, and human potential.

Amid this astonishing progress, an unsettling question remains: If aging is no longer simply a gradual descent into decline, how can we ensure these extra years are worthwhile?

This dilemma defines modern aging. Society venerates youth—equating it with ambition, vitality, and productivity—while often relegating older adults to the margins. Popular culture echoes this bias: youthful beauty is idolized, dynamic lifestyles glorified, and professional achievement prized above all else. Meanwhile, older people often confront workplace discrimination, social invisibility, and narratives framing aging as withdrawal rather than transformation. As longevity stretches further ahead, a greater challenge, a greater question emerges—what does it mean to age with intention.

Science, for all its wonders, can extend life, but it cannot instruct us on how best to live it. Here, philosophy steps into the silence, offering pathways toward wisdom and understanding. Existentialist thought, especially, has long confronted the question of living well in the face of life's fleeting nature. Unlike traditional thought that grounded life's purpose in external values—the Stoics' virtue, Aristotle's *eudaimonia*, or religious paths toward enlightenment—existentialism argues that meaning is never simply received; it must be actively and deliberately created.

Viktor Frankl, in *Man's Search for Meaning*, asserted that even in the darkest moments, we retain the capacity—and the responsibility—to choose meaning. Frankl, shaped by his experiences as a Holocaust survivor, did not see purpose as something fate hands us, but as something courageously constructed. In this light, aging becomes not loss, but an opportunity—an invitation toward deeper, deliberate living.

Jean-Paul Sartre took this idea further, famously declaring, "existence precedes essence." Human beings, Sartre insisted, arrive without predetermined identities or destinies; instead, each must actively shape their own purpose through thoughtful choice and authentic action. Rather than yielding to decline, aging thus becomes an ongoing process of self-definition—a continual act of reinvention and conscious engagement with existence.

Martin Heidegger expanded existential thought further still, emphasizing the crucial importance of confronting our mortality. Too often, he observed, we drift aimlessly—trapped in routines shaped by societal expectations rather than genuine aspirations, a disengagement that deepens as we age. Heidegger offered a powerful insight: fully acknowledging life's finite nature inspires us toward greater presence, clarity, and deliberate living. Aging, then, is not the passage of time but a summons toward authenticity—a call to focus on what matters most.

Modern society grants remarkable longevity—but longevity does not guarantee fulfillment. Intentional aging requires more than scientific innovation; it calls for a thoughtful philosophy that embraces rather than resists the passage of time, seeing later life not as a conclusion but as a continued exploration of potential. Perhaps this involves shifting focus from outward achievements toward inner satisfaction—deepening relationships, nurturing creativity, exploring lifelong passions, or mentoring younger generations. Whatever path is chosen, meaningful aging demands mindful and active participation in the world around us.

Today's extraordinary scientific advances encourage society to confront a fundamental practical and ethical challenge: building a culture that values the full arc of later life, ensuring aging remains a dignified, inclusive, and socially meaningful experience.

The Art of Aging: Philosophical Insights on Longevity and Identity

Contemporary philosophy offers a unique perspective on aging and longevity, exploring the ethical, social, and existential questions emerging from rapid advances in medical and technological innovation. As scientific progress deepens our understanding of human biology, philosophical inquiry invites reflection on how these developments may evolve the human experience. In modern culture, aging is often cast in a negative light—as something to resist, delay, or undo—a narrative reinforced by industries that prioritize perpetual youth. Philosophers challenge this narrow view, reframing aging as a meaningful stage of life rich with opportunities for personal growth, deepened relationships, and reflective self-discovery.

Two compelling modern philosophical orientations shed light on the complex issues surrounding aging. The first, often referred to as biotechnological optimism—or, in some interpretations, transhumanism—embraces the possibilities emerging at the frontier of science. Advocates highlight genetic research, regenerative technologies, personalized healthcare tools, and artificial intelligence as innovations that may expand how we live and care for ourselves. They envision scientific progress supporting autonomy, enabling older adults to maintain independence through tools that complement individual needs. This hopeful perspective must insist on ethical vigilance. Proponents emphasize the importance of ensuring fair access, dignity, and social justice so that technological advancements enhance well-being broadly rather than benefiting only a privileged few.

In contrast, bioconservatism—or existential caution—offers a reflective counterpoint. Supporters of this view encourage deeper contemplation of aging's intrinsic meaning. Life's finitude, they argue, brings authenticity, clarity, and emotional richness to human experience. Awareness of mortality can inspire gratitude, strengthen relationships, and heighten appreciation for life's fleeting beauty. However, this perspective cautions against overly aggressive efforts to alter aging through technological means. Such ambitions, they suggest, may introduce unintended social pressures or strain shared resources. In their view, aging invites us to move with life's natural rhythms rather than attempting to circumvent them, cultivating meaning through acceptance rather than control.

Both philosophical orientations reach into everyday realities such as caregiving and dependency. Philosophers examining aging highlight that dependency is neither a weakness to be "solved" by technology nor a limitation to be accepted passively. Instead, caregiving becomes an expression of human interconnectedness—an evolving relationship grounded in empathy, dignity, and mutual respect. From this standpoint, technological innovation is not opposed to ethical sensitivity; rather, the two can complement one another. Together, they support approaches to aging that honor both human capability and human vulnerability, enriching the lived experience rather than focusing solely on extending its duration.

Aging with Intention: Philosophy as a Practical Guide

Philosophy invites us beneath life's surface, uncovering deeper significance within the natural rhythms of aging. Drawing from this wisdom, we understand aging as more than the passage of years—rather, it unfolds as a journey toward authenticity and purposeful living. Stoic acceptance, Aristotelian flourishing, Taoist harmony, and existential courage each embody distinct forms of wisdom. Aging thus emerges not as a curse but as a path toward greater self-understanding.

Practically applying these philosophical principles enriches daily life. Older adults need not passively accept limiting narratives. Instead, pursuing activities that align closely with personal interests and capabilities can make aging vibrant and rewarding. Regular physical activity suited to individual abilities, mindful nutritional choices, intellectual exploration, and creative pursuits like art, music, or writing help transform aging into an inspiring life chapter.

Communities and policymakers play critical roles by fostering supportive environments. Strategic initiatives—such as universally accessible spaces, adaptive housing, intuitive technologies, and intergenerational mentorship—promote independence and dignity. Philosophical reflection also inspires advocacy for inclusive healthcare systems, equitable access to longevity treatments, culturally responsive healthcare services, and targeted outreach, ensuring respect and justice in later life.

Compassionate caregiving anchors these ideals, emphasizing autonomy, regard, and mutual understanding. Programs like respite care, caregiver education, and robust community networks offer essential support, allowing aging to become a time of creativity and meaningful contribution, recognized and genuinely valued by society.

Aging well involves crafting a life deeply engaged with others and aligned with personal values and aspirations. Philosophy empowers this process, transforming aging from something endured into an affirmation of our shared humanity—a testament to growth, strength of spirit, and attentive participation.

Aging well is both art and philosophy—an integration of reflective wisdom and deliberate choice. Beyond the passage of years, aging becomes an exploration of identity, purpose, and connection. Ancient philosophical traditions offer enduring guidance toward acceptance and insight, while contemporary thought reimagines aging as an intentional process of continued growth. At the intersection of long-standing wisdom and modern insight, aging emerges as a source of enrichment—inviting heightened self-awareness, deeper relationships, and experiences that carry meaning beyond the moment. It urges reflection on what constitutes genuine success, affirms the value of connection, and calls for a more attentive presence within each passing day.

CHAPTER 3

Why We Age: Scientific Theories on Life's Fundamental Mystery

"Youth is a perpetual intoxication; it is a fever of the mind. Aging is what leads us toward maturity, reflection, and deeper wisdom—it is life's own narrative of discovery."
—MILAN KUNDERA

Aging touches every living thing, yet it remains one of nature's most intricate and fascinating processes. All organisms age. It is a universal story, visible through physical transformations, gradual physiological shifts, and an increasing association with various health challenges. For centuries, scientists have grappled with a deceptively simple yet profound question: Why do we age? In seeking answers, numerous theories have emerged—each illuminating distinct facets of a beautifully complex interplay among biology, genetics, and environment.

No single theory fully captures the essence of aging. Instead, these ideas interlace, reinforcing one another, forming a richly textured narrative. Together, they help define our modern understanding of longevity, aging, and the delicate dance between biological stability and change.

This chapter explores the most influential theories, examining their underlying mechanisms and highlighting their interconnected contributions. By tracing these scientific pathways, we deepen our appreciation of aging—not as an inevitability, but as a complex, evolving process. When understood as a dynamic choreography of biology and experience, it shifts from something to dread into a human story rich with opportunity for insight, meaning, and growth.

The Damage Accumulation Theory: Aging as Biological Erosion

Picture an intricate, carefully maintained clockwork—a complex mechanism humming reliably for years. Gradually, tiny imperfections accumulate; gears lose their alignment, springs slowly weaken, and the system's precision inevitably begins to falter. Aging, according to the Damage Accumulation Theory, mirrors this slow erosion of function. Over time, cellular and molecular damage steadily accumulates, interacting with our body's remarkable capacity for self-repair and renewal in ways that modulate how these systems operate. Both internal metabolic activity and external environmental stressors act as relentless forces, contributing to biological change along converging pathways.

Central to this phenomenon are mitochondria, the cellular structures often celebrated as the body's microscopic power plants. These tiny organelles tirelessly produce energy essential for life, but there's a cost: energy production generates reactive oxygen species (ROS), unstable molecules akin to stray sparks escaping from a fire. Normally, the body's antioxidant defenses swiftly neutralize these sparks, maintaining a delicate equilibrium. However, with advancing age, shifts occur in these defenses, allowing oxidative stress—the imbalance favoring ROS—to interact more prominently with DNA, proteins, and cellular membranes. This cumulative oxidative burden contributes to declining cellular integrity, altering microscopic processes that can emerge over time as features of aging.

Internal mechanisms are not solely responsible. External environmental forces intensify this biological erosion. Consider ultraviolet (UV) radiation, encountered daily with every sunny stroll outdoors. UV rays continually bombard skin cells, triggering genetic alterations such as pyrimidine dimers, which disrupt DNA structure and affect pathways involved in cellular repair.

Similarly, airborne pollutants, including fine particulate matter (PM2.5), enter the respiratory system and interact with inflammatory signaling pathways. The lungs can be imagined as embattled landscapes, persistently exposed to microscopic irritants whose effects are well documented in both cardiovascular and respiratory biology.

Lifestyle choices further interact with these cumulative changes. Take smoking, for instance: every inhalation floods cells with compounds that heighten oxidative stress and molecular instability. Inside the body, proteins may occasionally misfold, forming tangled clusters within cells. Not every misfolded protein is harmful, but substantial aggregations—such as the amyloid plaques characteristic of Alzheimer's disease or the neurofibrillary tangles seen in Parkinson's disease—highlight the potential consequences of unchecked molecular accumulation.

Aging isn't only physical. Emotional experiences—once considered intangible and separate from biology—are now understood to interact with physiological pathways. Chronic psychological stress activates the hypothalamic-pituitary-adrenal (HPA) axis, flooding the bloodstream with cortisol, a stress hormone that influences inflammatory and metabolic signaling. Picture someone burdened by prolonged caregiving, financial insecurity, or loneliness, their cells continually bathed in stress-related biochemical cues. Over months and years, this persistent stress corresponds with biological aging markers such as telomere shortening—the gradual erosion of protective DNA caps—highlighting the intimate connections between emotional life and cellular expression.

However, even amid these accumulating influences, the body is far from helpless. Remarkably sophisticated defense systems work tirelessly behind the scenes, engaging continuously with biological change. DNA repair enzymes address altered genetic material, robust antioxidant networks interact with ROS, and cellular recycling processes—like autophagy—diligently clear away internal debris. Imagine these systems as formidable maintenance crews, continually tending to cellular well-being and preserving our internal landscape. But even the most diligent workers eventually shift in efficiency. With age, changes appear in how these protective mechanisms function as cumulative biological pressures increase. Over time, the interplay between repair and accumulation gives rise to hallmark features of aging: altered cellular activity, evolving tissue characteristics, and variability in organ performance.

Understanding this interplay underscores a critical insight: aging is neither purely biological nor isolated in its causes. Rather, it emerges from the dynamic interaction of genetic factors, metabolic processes, environmental exposures, and personal lifestyle decisions. Theories emphasizing oxidative damage, genetic instability, telomere dynamics, and cellular dysregulation do not stand apart; they form interconnected threads weaving a comprehensive, multidimensional picture of aging.

This interconnected understanding offers perspective. By being intentional about lifestyle—minimizing harmful environmental exposures, managing psychological stress, embracing nutritious diets, prioritizing regular physical activity, and supporting restorative sleep—people engage more deliberately with aging. Although we cannot pause time, sustained practices in daily life influence how a life is experienced.

Exploring these complex scientific accounts transforms our perception of aging. Viewing aging through the prism of cumulative damage and biological repair fosters a deeper respect for life's delicate balance, reminding us that while our choices influence our experience, they unfold within nature's inescapable rhythms and boundaries.

The Cellular Senescence Theory: The Persistent Shadow of "Zombie" Cells

Every thriving city depends on careful upkeep—damaged buildings swiftly repaired, those beyond saving replaced. But when vigilance falters, wear quietly spreads, altering the city's vitality piece by piece. Our bodies face a similar challenge through cellular senescence: damaged cells cease dividing yet remain in place, lingering silently within tissues. Early on, these so-called "zombie" cells function as protective guardians, halting the spread of potentially dangerous growths. With advancing age, however, their numbers tend to increase, and their presence shifts from helpful sentinel to quiet saboteur, influencing the tissues they once protected. Slowly and almost imperceptibly, these persistent cells move from guardianship to instigators of biological disruption, casting a long shadow across the body's internal landscape.

The influence of these tenacious cells emerges primarily from their secretion of a mixture known as the senescence-associated secretory phenotype (SASP). This inflammatory cocktail includes cytokines (molecules that ignite inflammation), chemokines (substances attracting inflammatory cells), and enzymes capable of interacting with structural proteins. To visualize SASP's impact, consider your own skin. Over time, senescent fibroblasts show reduced collagen production—collagen being crucial for skin firmness and elasticity. Worse yet, they release enzymes that interact with existing collagen and elastin in ways that produce visible changes such as wrinkles, sagging, and age spots. The mirror reflects these molecular realities: youthful elasticity gives way to deepening lines and lost firmness.

This pattern appears in many tissues. In arteries, SASP-related signaling drives vascular stiffening and shifts in circulatory dynamics that affect cardiovascular function. Yet the disruption goes even deeper. As understanding has advanced, a striking revelation has emerged: senescent cells do not remain passive nuisances; they actively eject fragments of damaged mitochondria—particularly mitochondrial DNA and double-stranded RNA—into surrounding tissues. Like tiny distress signals, these fragments engage innate immune pathways, contributing to chronic, low-grade inflammation, a hallmark of aging.

Alarmingly, senescent cells can propagate their disruptive influence. Under certain conditions, their inflammatory signals prompt neighboring healthy cells to enter senescence themselves, effectively recruiting new members into the ranks of altered cellular behavior. Although the speed and extent of this process vary across different tissues, its presence is unmistakable—a self-sustaining cycle of inflammation, cellular wear, and cumulative burden.

Over time, the accumulation of zombie cells coincides with shifts in immune function. The body's natural defenses lose efficiency, influencing how effectively they clear cellular debris, respond to challenges, or maintain balance. Consider the brain, where senescent glial cells—accompanied by chronically activated microglia and astrocytes—foster persistent neuroinflammatory states. Like static interfering with a clear radio signal, this inflammation can disrupt neural communication in ways seen in cognitive aging.

Cellular senescence embodies a striking biological paradox: beneficial during youth yet increasingly burdensome with age. Recognizing this duality opens new lines of inquiry. Scientists are now exploring whether altering, modulating, or removing senescent cells—or neutralizing their signals—might influence biological aging.

If these cells are envisioned as neglected buildings threatening an otherwise thriving biological city, the challenge becomes one of careful intervention: developing strategies to address them safely, while preserving stability within the surrounding cellular neighborhood.

The Cellular Impact of Glycation: Aging's Hidden Culprit

Think of a well-used kitchen sponge—initially flexible, absorbent, and resilient. Over time, however, repeated exposure to moisture, heat, and daily wear leaves it stiff, brittle, and less effective. Similarly, our bodies experience an internal transformation through a process called glycation, slowly changing tissues from supple and flexible to stiffened and fragile. Glycation occurs when sugars chemically bind to proteins or fats, forming molecular complexes known as advanced glycation end-products (AGEs). These molecular "adhesives" accumulate, interacting with cellular structures and influencing tissue behavior in ways that mirror the biological changes of aging.

The effects of glycation are both visibly apparent and internally disruptive. Externally, the skin provides a clear example. AGEs accumulate within the collagen fibers that grant youthful skin its supple elasticity, gradually stiffening these fibers until skin becomes less flexible and resilient. Picture someone noticing deepening wrinkles or sagging skin, not as superficial signs of age, but as outward signals of internal molecular stiffening. Beneath the surface, these once-flexible protein structures—collagen and elastin—lose elasticity, appearing externally as familiar signs of aging.

Internally, the impact extends deeper. Within arteries, glycation's molecular stickiness affects the flexibility of vessel walls. As arteries stiffen, the smooth movement of blood can be altered. Over the years, this gradual progression can place greater workload on the heart, resulting in hemodynamic alterations characteristic of vascular aging. Much like pipes in an aging plumbing system becoming clogged and brittle, stiffened arteries alter blood flow and oxygen delivery, leaving a distinct imprint on the body's circulatory landscape.

In the brain, glycation operates like an unseen saboteur, contributing to neurodegenerative change. AGEs promote the formation and stabilization of protein clumps such as amyloid plaques and tau tangles—hallmarks of Alzheimer's disease. Think of these protein aggregates as tangled threads within a delicate tapestry, interrupting neural communication pathways and gradually affecting cognitive clarity. Over decades, cumulative glycation correlates with shifts in memory, clarity, and cognitive agility, underscoring its deeper, insidious influence.

However, the rate and extent of glycation appear responsive to daily choices. Elevated blood sugar levels—often associated with diets high in refined carbohydrates and sugars—accelerate AGE formation. Each sugary snack or highly processed meal can influence microscopic processes connected to the biology of aging. Fortunately, practical, intentional dietary adjustments may help reduce glycation pressure. Reducing refined sugars, choosing balanced nutrition rich in antioxidants, and supporting stable blood glucose levels limit AGE accumulation and help preserve tissue structure. These choices, made consistently, align with pathways that favor healthier aging.

Scientists, equipped with deeper insights into glycation's molecular mechanisms, continue to investigate ways to better understand its effects. Current research explores approaches ranging from pharmaceutical compounds that interfere with AGE formation to lifestyle-based strategies that support the body's intrinsic capacity to manage molecular damage. Could modulating glycation represent a meaningful step in altering how aging unfolds—shifting it from a passive inevitability to a process that can be observed and understood more clearly? It remains an intriguing question.

Understanding glycation offers perspective, shifting our understanding of aging from a passive experience of decline into one marked by greater awareness. With this deeper knowledge, aging no longer signifies the ticking clock of biological fate; rather, it becomes a story shaped by attention—a process enriched by deliberate, informed decisions. Recognizing glycation's inconspicuous yet widespread effects underscores the importance of continued scientific research—not solely for groundbreaking discoveries, but because even incremental advances can contribute to improvements in how people experience health and quality of life as they age.

The Genetic and Epigenetic Theories of Aging: Decoding Longevity's Blueprint and Dynamic Expression

Within every living cell lies DNA, an elegant molecular blueprint that orchestrates life's most fundamental processes—growth and healing, adaptation and resilience. According to the the Genetic Theory of Aging, certain genes linked to longevity serve as biological caretakers, diligently contributing to cellular maintenance, supporting metabolic balance, and helping the body respond to stress. Genes never operate in isolation; instead, they participate in intricate networks, collectively moulding our body's capacity to repair itself, sustain equilibrium, and navigate the accumulating wear of time.

Genetic inheritance varies considerably from one person to another, deeply influencing each person's distinctive path of aging. Some people inherit beneficial genetic variants—biological blessings connected to more efficient DNA repair, refined metabolic tuning, and stronger responses to stressors. Others carry genetic vulnerabilities: mutations or less favorable variants tied to reduced repair capacity or altered stress-response pathways, contributing to changes commonly seen in age-related decline. But genes alone never write our aging story. Instead, aging emerges through a dynamic dialogue between inherited potential and external influences. Choices involving nutrition, physical activity, stress management, and environmental exposures contribute to how genetic tendencies unfold, affecting the degree to which we realize the longevity potential encoded in our DNA.

Building upon this genetic foundation, the Epigenetic Theory of Aging offers a more nuanced perspective on how gene expression dynamically evolves throughout a lifetime. If DNA serves as life's foundational script, then epigenetics acts as an interpreter—carefully deciding when genes awaken or fall silent. Epigenetic regulation occurs through chemical modifications, primarily DNA methylation (chemical tags that attach to DNA and guide gene activity) and histone acetylation (which influences how tightly DNA is wrapped and thus how accessible genes become). These chemical adjustments do not alter the genetic code itself; instead, they finely tune gene expression in response to internal physiological states, daily behaviors, and environmental exposures.

Over decades this precise epigenetic choreography can shift, leading to what scientists call "epigenetic drift." Beneficial genes involved in cellular repair, inflammatory balance, and metabolic efficiency may become less active, while other genes tied to inflammatory or tissue-altering pathways may become more active. Gradually, this loss of epigenetic precision coincides with changes in cellular function, shifts in immune activity, and reduced regenerative capacity—features often recognized as hallmarks of biological aging unfolding silently beneath the surface.

Importantly, epigenetic drift is neither fixed nor fully predetermined. Because epigenetic patterns remain inherently responsive, they offer intriguing avenues for exploring how aging can be influenced. Current scientific work examines ways to adjust or reset these molecular markers, investigating whether aspects of age-related biology can be modulated. Could future interventions restore elements of youthful epigenetic precision, altering the course of aging in meaningful ways? It is a question that continues to energize researchers.

At the forefront of these developments stand epigenetic clocks—innovative tools that estimate biological age by analyzing patterns of DNA methylation. Unlike chronological age, which simply counts birthdays, biological age reflects characteristics tied to physiological status at the cellular level. Two people with identical chronological ages can display strikingly different biological profiles, shaped by genetics, diet quality, physical activity, stress levels, and environmental exposures. Epigenetic clocks interpret these molecular fingerprints, offering a personalized snapshot of biological aging.

The potential implications of epigenetic clocks are expansive. By revealing molecular indicators that reflect accelerated or slower biological aging, scientists can evaluate which lifestyle choices or research-based interventions align with more favorable outcomes. In this way, epigenetic clocks serve not just as measurements but as tools that help illuminate how behaviors and exposures correspond to biological aging.

Advances in non-invasive epigenetic testing have expanded access to biological age assessments. Simple techniques, such as cheek swabs or saliva samples, now allow regular tracking of epigenetic markers without the need for invasive procedures. This convenience supports longitudinal research into molecular aging over time, enabling frequent, real-world monitoring across extended periods.

Moreover, innovative tissue-specific epigenetic clocks offer even greater precision, revealing distinct aging signatures across different tissues and organs. Cells in the liver, brain, heart, and immune system may age at differing rates, underscoring the intricate variability of the aging process. Recognizing these differences allows researchers to explore strategies tailored to particular tissues—whether examining cognitive resilience, cardiovascular function, or immune performance.

Uncovering the intricate dance between genetics and epigenetics transforms our fundamental understanding of aging. No longer viewed as predetermined genetic fate nor solely environmental consequence, aging emerges as a vibrant, ongoing dialogue between inherited potential and daily experience. This rich interplay explains how two people with similar genetic backgrounds can follow dramatically different aging journeys. Every choice—dietary habits, physical activity, stress management, sleep quality, and social connection—sends epigenetic signals that influence gene expression in ways that accumulate across a lifetime.

Integrating these insights changes how aging is understood. Longevity becomes not merely inherited potential but an invitation to participate meaningfully—a narrative formed through awareness, intentional action, and the evolving science of epigenetics. Aging shifts from a story of inevitability to one of relationship: a dynamic exchange between biology and experience, marked by resilience, adaptation, and human depth.

Aging, in short, is not destiny. It is dialogue.

The Mitochondrial Theory: Aging's Energy Crisis

At the heart of every living cell are mitochondria—tiny, tireless structures often described as cellular powerhouses. Like miniature factories humming constantly, they produce adenosine triphosphate (ATP), the indispensable energy currency that fuels virtually all biological activities. Yet this continuous energy production comes at an inevitable price. While generating ATP, mitochondria simultaneously produce reactive oxygen species (ROS)—highly unstable molecules that behave like microscopic sparks, capable of interacting with critical cellular components such as proteins, lipid membranes, and mitochondrial DNA (mtDNA).

Early in life, antioxidant defenses efficiently neutralize reactive oxygen species (ROS), limiting cellular damage and helping maintain internal balance. Over time, the capacity of these defenses often declines, allowing oxidative reactions to accumulate. Mitochondrial DNA is particularly susceptible: located near sites of ROS generation and supported by relatively limited repair mechanisms, it gradually accrues molecular damage. This accumulation can contribute to a reinforcing cycle in which compromised mitochondria generate less ATP while producing higher levels of ROS, with downstream effects on cellular performance and stability.

These mitochondrial shifts are most evident in energy-demanding tissues. Consider the brain, a dynamic organ constantly craving fuel. Here, mitochondrial dysfunction can alter neural activity in ways that overlap with what is known about cognitive aging and increased vulnerability to neurodegenerative conditions such as Alzheimer's and Parkinson's disease. Within the heart—a muscle tirelessly contracting billions of times throughout a lifetime—reduced mitochondrial efficiency may influence cardiac performance and contribute to physiological changes seen in cardiovascular disease. Skeletal muscles, too, reflect mitochondrial decline: sarcopenia, the gradual age-related loss of muscle mass and strength, corresponds with changes in mitochondrial performance that can affect mobility, independence, and overall quality of life.

Mitochondria do far more than produce energy. They also participate in cell signaling, regulate programmed cell death (apoptosis), and contribute to oxidative-stress management. As mitochondrial activity declines, these regulatory roles can shift as well, influencing cellular balance and affecting how aging manifests across interconnected systems.

Given their central role in aging biology, mitochondria have become a major focus of longevity research. Scientists are actively exploring targeted strategies—from caloric restriction, fasting, and mitochondrial-specific antioxidants to innovative pharmacological and biotechnological approaches—aimed at supporting mitochondrial stability. Research continues to examine how lifestyle behaviors interact with these strategies, exploring their relevance to longevity-related outcomes. Although extending lifespan remains a complex scientific question, efforts to support mitochondrial function reflect a broader goal: promoting resilience, maintaining capability, and sustaining independence across the added years that longer lives provide. Aging thus becomes not a story of decline, but an engaging scientific and personal quest—one that encourages us to understand, nurture, and support the cellular foundations of life.

The Inflammation Theory: Aging's Quiet Fire

Beneath the calm surface of the body simmers a silent yet persistent fire—one faint enough to remain unnoticed, yet fierce enough to influence health over decades. Scientists have termed this chronic internal smoldering "inflammaging," a condition characterized by steady, low-level inflammation deeply embedded in the aging process. Unlike acute inflammation—the body's rapid, protective response to injury or infection—inflammaging chronically endures, placing continuous demands on cells and tissues and gradually influencing biological resilience.

Multiple forces feed this fire. Central among them are senescent cells, those biological "zombies" that no longer divide but stubbornly persist. Though dormant, these cells are hardly silent. As previously discussed, they secrete an inflammatory mix known as the senescence-associated secretory phenotype (SASP)—a biochemical brew of molecules that acts like signals leaping to surrounding healthy cells, encouraging inflammatory responses. Through these signaling molecules, senescent cells can influence neighboring cells toward similar states, contributing to a self-sustaining cycle of cellular distress.

Oxidative stress further fuels the flames. Reactive oxygen species (ROS), described earlier in mitochondrial function, interact directly with inflammatory signaling pathways. These oxidative interactions accumulate steadily, prompting immune-cell activation and intensifying the inflammatory cycle. The body finds itself in a constant internal negotiation, caught in a loop in which cellular damage and inflammatory signaling reinforce one another.

Even the gut microbiome—the vibrant community of microorganisms residing within the digestive tract—plays a crucial role. With age, the once-diverse microbiome may shift toward imbalance, moving away from beneficial microbes and toward species that promote inflammatory states. Concurrently, the intestinal barrier can become more permeable, a condition commonly referred to as "leaky gut." This porous lining allows microbial components to enter the bloodstream, prompting an ongoing, low-grade immune response that contributes to systemic inflammation.

The consequences ripple widely, affecting organs from skin to brain. Within arteries, persistent inflammation can alter vessel-wall structure and contribute to physiological changes characteristic of cardiovascular disease. In the brain, chronic inflammatory states can disrupt neural function and increase vulnerability to cognitive decline and neurodegenerative disease. Even the immune system itself can shift over time, its vigilance gradually altered by decades of low-level activation and increasingly variable responsiveness.

With greater understanding comes the possibility of more informed engagement. Researchers are examining multiple ways to moderate inflammaging, including diets rich in antioxidants and omega-3 fatty acids, as well as lifestyle practices such as regular physical activity, meditation, and restorative sleep. Pharmacological approaches targeting inflammatory pathways are also under investigation, alongside emerging lines of inquiry involving microbiome modulation and senolytic strategies. Taken together, these efforts reflect an expanding scientific attempt to better characterize—and in some cases influence—the biological processes that sustain chronic inflammation with age.

In embracing these scientific insights, aging shifts from passive inevitability to a more deliberate process. Rather than accepting the quiet inflammation smoldering within, people can engage in behaviors known to interact with inflammatory biology. Understanding the underlying mechanisms of inflammaging presents aging as a dynamic conversation between biology and experience, revealing opportunities for thoughtful engagement with long-term well-being.

The Telomere Shortening Theory: Aging's Cellular Countdown

Within each cell resides DNA—life's precious instruction manual—protected at its ends by specialized molecular "caps" known as telomeres. Often likened to the plastic tips on shoelaces that prevent fraying, telomeres steadily shorten each time a cell divides, gradually reducing their protective capacity. Eventually, telomeres reach a critical length, signaling cells that their healthy lifespan is nearing completion. At this crossroads, cells either enter senescence—a dormant state reminiscent of cellular zombies—or undergo apoptosis, a precisely choreographed form of programmed self-destruction.

Telomere dynamics reveal themselves vividly in tissues characterized by rapid and continual regeneration, such as the immune system and digestive tract. Consider immune cells—essential defenders against biological challenges. Early in life, ample telomere length supports strong proliferative responses. As years accumulate and telomeres shorten, shifts can emerge in immune-cell renewal, influencing how these cells respond to stressors and microbial encounters. Similarly, the lining of the gut—dependent on constant renewal to sustain digestive processes—may exhibit structural and functional changes as telomeres shorten. In both cases, internal cellular aging can alter functional capacity, subtly impacting vitality from within.

Still, telomeres do more than record time's passage—they form part of the molecular system that reflects how aging unfolds. Intriguingly, daily experiences appear to influence telomere behavior. Chronic stress, diets rich in refined sugars, and sedentary lifestyles correspond with faster telomere shortening, whereas balanced nutrition, regular physical activity, intentional stress management, and supportive social connection correspond with slower telomere loss. These relationships illustrate how lived experience and cellular biology continually communicate with one another.

Remarkably, researchers have begun building on this insight, exploring interventions intended to support telomere stability or influence telomerase activity (the enzyme involved in rebuilding telomeres). Current work—including pharmacological investigations, lifestyle-oriented studies, and novel molecular approaches—seeks to clarify how telomere maintenance relates to biological aging. These lines of inquiry raise compelling questions: Could modulating telomere dynamics alter the course of age-related change? Might such strategies affect how tissues respond as aging progresses? Scientific inquiry continues to unfold.

Understanding telomere biology reframes aging from passive inevitability into a more participatory trajectory. Rather than only marking the passage of biological time, telomeres serve as dynamic indicators—reflecting the interplay among genetics, environmental influences, and lived experience. Studying these molecular structures reveals the constant negotiation between inherent biological limits and emerging scientific possibilities that may one day help clarify, and perhaps extend, the boundaries of cellular life.

The Neuroendocrine Theory: Aging's Hormonal Symphony

At the core of aging lies a delicate biological symphony—a harmonious interplay of hormones masterfully conducted by the brain's hypothalamus. In youth, this intricate hormonal orchestra performs with remarkable precision, synchronizing metabolism, growth, stress responses, reproduction, and tissue repair into a unified, resilient whole. Over decades, the precision of this hormonal symphony fades, gradually slipping into imbalance. Melodies that once flowed effortlessly become discordant, contributing to the complex patterns of aging.

Consider growth hormone, a key player in tissue maintenance, muscle vitality, and skin structure. In our younger years, abundant growth hormone supports robust tissues—muscles recover swiftly, skin remains firm, and bones maintain their strength. But with aging, its production diminishes like a slowly fading musical note, and the body's regenerative capacities shift. Muscles may feel less vigorous, wounds take longer to heal, skin becomes thinner and more delicate, and bones gradually lose density.

Similarly, estrogen and testosterone—vital reproductive hormones—decline over time, casting their own shadows over lived experience. As these hormones ebb, people often report shifts in cognitive sharpness, changes in bone density, and alterations in cardiovascular function. The steady hormonal harmony of youth gradually gives way to imbalance, influencing the body's internal balance from within.

Adding complexity to this hormonal symphony is cortisol, the body's primary stress hormone. Normally, cortisol serves as a powerful protector in short bursts, preparing us for action or alertness. But as we age, cortisol levels can remain chronically elevated, a once-helpful melody now echoing continuously through the biological orchestra. Persistently high cortisol drives increased inflammatory signaling, alters immune responsiveness, and affects cognitive clarity—transforming a beneficial short-term response into a long-standing physiological challenge.

This gradual unraveling of hormonal harmony has inspired significant scientific inquiry. Researchers continue to investigate how hormonal changes interact with lifestyle variables such as balanced nutrition, regular physical activity, mindful stress management, and quality sleep. These factors interact with hormonal dynamics across the lifespan. Related lines of investigation also examine hormone replacement strategies, weighing potential benefits, risks, and the contexts in which such approaches may be considered under medical guidance. Rather than reversing aging outright, these efforts reflect attempts to understand how hormonal balance contributes to well-being.

This perspective encourages a more deliberate engagement with aging. By understanding the body's shifting hormonal dynamics, people can participate intentionally in supporting physiological balance. Aging becomes not a story of decline, but a dynamic process informed by awareness, thoughtful choices, and scientific insight. It invites recognition of the complexity of hormonal interactions and reminds us that sustaining well-being with age depends not on pursuing eternal youth, but on maintaining balance within the body's interconnected biological systems.

The Disposable Soma Theory: Evolution's Delicate Balancing Act

Why would evolution, so tirelessly focused on refining life for survival, allow organisms to age? The Disposable Soma Theory provides an insightful answer: aging isn't a biological oversight but a carefully calibrated compromise shaped by resource constraints. Life faces a perpetual balancing act between allocating limited energy resources toward reproduction—the essential task of passing genes to future generations—and preserving bodily integrity (the soma) through continuous maintenance and repair.

This evolutionary bargain plays out vividly across the natural world. Take mice, for instance—tiny creatures racing against time, investing furiously in rapid reproduction. They produce multiple litters in quick succession, channeling their energy into offspring rather than long-term self-preservation. The consequence? Their lifespan is fleeting, a brief spark that fades quickly once reproductive tasks are complete. Humans, whales, and other long-lived species, by contrast, initially direct substantial energy toward maintaining bodily systems, allowing extended reproductive opportunities and better odds of survival. However, even among these species, this commitment to cellular upkeep eventually shifts. As reproductive potential wanes, evolutionary pressures on maintenance lessen, and aging becomes more pronounced.

Viewing aging as this evolutionary negotiation transforms how we understand our biological journey. Aging isn't nature's flaw—it's a logical consequence of life's pragmatic strategy to favor reproduction and genetic legacy over indefinite physical upkeep. Recognizing the rules of this ancient compromise offers a sense of perspective: if resource allocation influences aging, it becomes possible to examine how behaviors interact with these biological tendencies.

Practical actions—including intentional dietary choices, regular physical activity, deliberate stress reduction, and continued scientific exploration—relate to cellular maintenance and metabolic balance. By engaging in practices that support repair processes, the experience of aging shifts from passive observation toward active participation. Even as such strategies are adopted, aging remains inherently complex, conditioned not only by individual decisions but also by genetics, environmental influences, and the deep evolutionary trade-offs embedded within life.

Mosaic of Time: Weaving Biology, Complexity, and Purpose

Aging is not just a biological progression or a passive drift toward inevitable frailty; it is a complex experience artfully pieced into the mosaic of human existence—a meaningful and multidimensional journey. During youth, sophisticated biological mechanisms support vitality, contributing to cellular stability and overall well-being. Over time, however, imperceptible shifts occur; protective processes gradually change, tipping the body's balance toward greater vulnerability.

Genetic blueprints, epigenetic modifications, metabolic rhythms, and cellular defenses do not operate in isolation; instead, they continually interact, influencing one another in dynamic ways. Aging can be envisioned as an intricate ecosystem, composed of diverse biological elements—each playing an essential role yet continually adjusting as the years advance. Individually, these elements hold significance; collectively, their interactions reveal the complexity of aging and the gradual transformations unfolding over time.

This biological dynamic remains highly responsive to lived conditions. Aging is not a fixed destiny dictated solely by genetics; it follows an evolving trajectory influenced by how life is lived. While aging cannot be halted, its course is shaped by familiar elements of daily life: nutrition, movement, stress, rest, and engagement with emerging scientific insight.

With this understanding, aging shifts from passive acceptance to a more engaged and deliberate way of living. It encourages steadiness amid change, meaningful connection with others, and reflection on what gives life its deeper purpose. Aging becomes a dialogue between biology and choice, where the ways we meet each day—how we care for ourselves and how we engage with the world—help define the experience of growing older.

Part II
Foundations of Longevity— Core Strategies for Health and Vitality

Aging represents more than the passage of years; it is the ongoing refinement of life—a dynamic journey defined by daily choices and actions. Modern research suggests that aging is influenced by how people move, eat, rest, manage stress, and connect with others. Each choice contributes to the broader course of our well-being.

Movement, for example, reflects autonomy and vitality. Beyond exercise, regular physical engagement supports muscular, skeletal, and cognitive functions, reinforcing the capacity for an active life. Nutrition likewise is more than basic sustenance; it can influence cellular processes, energy availability, and overall physiological balance. Sleep—often underestimated—contributes to mental clarity, immune regulation, and emotional steadiness. Stress-management practices complement these foundations by helping us navigate uncertainty with greater psychological and physiological composure.

Genuine longevity transcends the physical alone. Comprehensive wellness embraces emotional adaptability, spiritual enrichment, and the deep value of meaningful relationships. Bonds of friendship, companionship, and intimacy create purpose and depth, enriching our experience of aging. Through intentional connection, growing older becomes not just continued existence but an expanding source of meaning.

The chapters ahead move from broad principles to practical pathways—exploring how movement supports physical capability, how nutrition sustains overall function, and how mindful decisions and emotional well-being influence the quality of daily life. Seen in this way, aging becomes less a predetermined decline and more an unfolding narrative guided by awareness and intention.

By embracing this perspective, a clearer sense of direction emerges in the experience of growing older. Each year becomes not something to endure, but an opportunity—an invitation to cultivate strength, deepen emotional steadiness, and engage more fully with one another. Perhaps the greatest reward of aging lies not in orchestrating every moment but in recognizing the quieter rhythms of life—the unspoken moments that connect us in our shared humanity.

CHAPTER 4

Movement and Longevity: Motion Across a Lifetime

"The journey of a thousand miles begins with a single step."
—Lao Tzu

One practice holds extraordinary power to strengthen the heart, sharpen the mind, and extend both healthspan and longevity—movement. Far more fundamental than a fitness tool or transient health trend, regular physical activity is a biological imperative embedded within our physiology. At the cellular level, movement initiates coordinated adaptive responses.

Exercise goes well beyond calorie burning and muscle building; it activates potent cellular repair mechanisms, regulates inflammation, and fortifies critical systems in ways that influence how aging unfolds. Maintaining mobility, flexibility, and strength ensures that aging need not equate to weakness. Instead, it can mean preserving autonomy, safeguarding physical capability, and fully participating in everyday activities. Each walk, stretch, or resistance-training session reinforces resilience, supports functional capacity, and enriches quality of life.

Its influence surpasses the physical, benefiting mental and emotional well-being. Physical activity enhances cognitive function and promotes neuroplasticity—the brain's remarkable ability to adapt and reorganize—processes relevant to neurodegenerative change. An active body supports an active mind. Those who move regularly experience greater clarity, creativity, and emotional stability.

Extensive scientific evidence shows that consistent movement is associated with improved cardiovascular health, stronger bones and joints, and better metabolic function. These changes correspond with more favorable health trajectories. Movement does more than reduce risk—it helps preserve the functional capacity that makes those trajectories livable.

Movement can be incorporated into daily life in many forms. Whether taking stairs instead of elevators, walking in nature, gardening, or engaging in basic strength-building exercises, every motion counts. Movement is more than a medical intervention—it fosters freedom, builds durability, and underpins an active, independent life.

The Evolutionary Perspective—Why Humans Are Designed for Movement

Human beings are not only capable of movement; we are biologically designed for it. Unlike species whose survival relies on brief bursts of speed or raw power, humans evolved predominantly for endurance. Our ancestors did not chase prey with rapid sprints; instead, they utilized "persistence hunting," patiently tracking animals over vast distances until exhaustion left prey unable to escape. This sustained physical effort shaped our physiology profoundly—from cardiovascular efficiency and muscular structure to metabolic flexibility.

Human anatomy reflects these adaptations clearly. Long limbs, upright posture, and efficient sweat glands for cooling enable sustained physical exertion without overheating. Predominantly slow-twitch muscle fibers conserve energy, enabling stamina over extended periods. Collectively, these evolutionary traits underscore a fundamental truth: consistent movement is essential to normal human physiological function.

This relationship between habitual movement and longevity is evident in contemporary examples of exceptional aging. "Blue Zones"—regions with notably high concentrations of centenarians—demonstrate how routine, integrated physical activity contributes to sustained vitality. Populations in places like Okinawa, Japan, and Sardinia, Italy, embed movement into daily life through long walks, gardening, and manual labor. This continuous activity preserves musculoskeletal and cardiovascular function while corresponding with lower rates of chronic disease.

Modern lifestyles, however, often distance us from our evolutionary blueprint. Prolonged sitting, reliance on motorized transportation, and reduced physical labor have disrupted our natural connection to movement, shifting metabolic and cardiovascular function in unfavorable directions and contributing to physical decline. Yet research consistently affirms that incorporating regular, moderate activity—ranging from leisure pursuits to structured fitness routines—can help restore physiological balance and favor healthier aging.

Movement, therefore, is not simply a lifestyle choice; it is a biological necessity for maintaining physiological integrity.

The Science of Movement: How Exercise Fuels Longevity

Exercise extends beyond physical fitness; it acts as a powerful biological catalyst, initiating cellular adaptations that significantly influence longevity. By enhancing energy production, genetic regulation, and cellular resilience, movement directly impacts core elements of the aging process, shaping physiological trajectories and extending healthspan.

One primary way exercise supports longevity is through mitochondrial biogenesis—the creation of new mitochondria, cellular "powerhouses" that produce energy (ATP). Central to this process is AMPK (adenosine monophosphate-activated protein kinase), an enzyme that senses cellular energy availability. During exercise, temporary energy depletion activates AMPK, prompting mitochondrial production to restore energy equilibrium. Over time, mitochondria naturally lose effectiveness, contributing to fatigue and greater vulnerability to age-related functional decline. Regular physical activity counters this decline, bolstering mitochondrial efficiency, improving glucose metabolism, enhancing insulin sensitivity, and influencing metabolic pathways implicated in diabetes, cardiovascular disease, and neurodegenerative disorders.

Aging also involves balancing tissue growth and repair—a delicate equilibrium responsive to exercise. Physical activity modulates two critical yet opposing biological pathways: mTOR (mechanistic target of rapamycin), a central cellular pathway that regulates growth, repair, and protein synthesis, and autophagy, the body's internal recycling system that degrades and removes damaged proteins, organelles, and other cellular components, supporting cellular renewal and metabolic homeostasis. Resistance training activates mTOR, stimulating muscle growth, tissue repair, and protein synthesis—processes vital for preserving muscle mass and strength, both of which decline with age. However, excessive nutrient availability and inactivity can cause mTOR to remain persistently active, suppressing autophagy. Aerobic exercise and intermittent fasting stimulate autophagy, promoting cellular renewal and limiting the accumulation of damaged proteins associated with neurodegenerative change. Thus, regular movement preserves the body's flexibility, resilience, and regenerative capacity.

Additionally, exercise preserves cellular integrity through proteostasis—the maintenance of healthy protein function within cells. Misfolded or damaged proteins can accumulate, disrupting cellular health and accelerating aging. Exercise activates heat shock proteins, molecular "chaperones" that repair or clear misfolded proteins before harmful aggregations form. This mechanism is particularly important for maintaining protein homeostasis in tissues vulnerable to age-related degeneration.

Physical activity further influences gene expression through epigenetic mechanisms. Though the genetic code remains stable, exercise fine-tunes gene activity via DNA methylation and histone modification. These alterations activate protective genes and silence harmful ones, modulating inflammation, oxidative stress, and metabolic regulation—key drivers of aging.

At its core, movement is more than exercise—it is a biological inheritance, encoded through millennia of adaptation. When we engage this primal rhythm, we go beyond preserving function; we alter the conditions under which aging unfolds, exchanging drift for direction and passivity for purposeful motion.

Inflammation and Oxidative Stress: Exercise as a Natural Regulator

As explored in depth in a previous chapter, chronic low-grade inflammation accelerates aging, fueling a cascade of physiological changes spanning cardiovascular, musculoskeletal, neurological, and metabolic systems. Although acute inflammation is beneficial and necessary—swiftly healing wounds such as a scraped knee through precise immune responses—persistent inflammation acts more like a smoldering fire, steadily eroding cellular health and leaving tissues increasingly vulnerable to age-related dysfunction.

Fortunately, regular physical activity serves as a powerful natural countermeasure by recalibrating the body's inflammatory signaling pathways. Exercise significantly reduces harmful pro-inflammatory cytokines, such as tumor necrosis factor-alpha (TNF-α) and interleukin-6 (IL-6), while simultaneously enhancing beneficial anti-inflammatory mediators, notably interleukin-10 (IL-10). This balance curtails chronic tissue stress, boosts immune efficiency, and fortifies systemic stability in the body's response to inflammatory burden.

In addition to its anti-inflammatory effects, regular physical activity strengthens the body's capacity to manage oxidative stress through adaptive signaling rather than simple suppression. During exercise, metabolic demand temporarily increases the production of reactive oxygen species (ROS). Although excessive ROS can damage cellular components, these transient elevations also serve as signaling cues that activate protective pathways. Repeated exposure to this controlled stress stimulates the up-regulation of endogenous antioxidant systems, including enzymes such as superoxide dismutase and components of the glutathione redox cycle. Over time, these adaptations improve redox balance and increase tolerance to oxidative challenges. In this way, moderate, repeated exercise promotes cellular adaptive capacity by conditioning biological systems to respond more effectively to oxidative load.

This interplay between inflammation and oxidative stress contributes to cellular aging by promoting the shortening of telomeres, which protect chromosome ends from damage. With each cell division, telomeres erode, eventually driving cells into senescence. Regular exercise moderates this attrition by attenuating inflammatory and oxidative conditions linked to telomere dynamics, preserving cellular integrity and sustaining the body's reparative capacity.

Moreover, exercise engages the body's regenerative mechanisms by activating tissue-specific stem cells. Satellite cells nestled within muscles mobilize with physical exertion, facilitating muscle repair and regeneration. Similarly, neural progenitor cells within the brain respond to regular movement, promoting neuroplasticity and sustaining tissue renewal. By nurturing these vital stem cell populations, consistent physical activity supports tissue maintenance and adaptive capacity.

Exercise cannot be reduced to isolated effects—it coordinates an expansive network of biological processes that influence the pace and character of aging. By optimizing mitochondrial efficiency, harmonizing cellular growth and repair, protecting telomere integrity, managing inflammation, and encouraging cellular renewal, physical activity emerges as an integrative influence on aging biology. Its combined effects reinforce the body's structural coherence, modulating how time's pressures are expressed within living tissues.

Exercise and Whole-Body Health: Strengthening Every System

Exercise is one of the most potent and versatile determinants of long-term physiological function, benefiting nearly every major system in the body.

By enhancing resilience, restoring internal balance, and shaping trajectories associated with chronic physiological strain, regular physical activity serves as a foundational catalyst for systemic function.

Unlike pharmaceutical approaches, which often isolate single diseases or symptoms, exercise engages a broad constellation of interconnected processes. Movement fortifies the heart, optimizes metabolic function, reinforces immune defenses, and guides cognitive function, making it a universally accessible, life-affirming influence on aging biology. Consistent physical activity creates a biological environment in which each system can perform optimally—allowing movement to become a central contributor to preserving functional capacity and biological integration.

The Heart and Circulatory System

A robust cardiovascular system stands as the cornerstone of longevity, and exercise plays a foundational role in its maintenance. Each episode of physical exertion places a controlled demand on the heart, and over time this repeated stimulus strengthens cardiac muscle, enhancing its capacity to pump blood efficiently throughout the body. This repeated stimulation lowers blood pressure, improves arterial elasticity, and is associated with more favorable cardiovascular function, including patterns linked to reduced vascular strain. Imagine arteries as flexible garden hoses, effortlessly transporting nutrients and oxygen; exercise helps keep these blood vessels supple, responsive, and less prone to structural stiffening.

Even modest amounts of physical activity yield substantial cardiovascular benefits. Short bouts of vigorous movement—such as brisk walking, energetic cycling, or focused resistance training—are associated with lower cardiovascular risk. By improving cardiovascular efficiency, regular movement allows tissues to receive nutrients and oxygen more effectively, supporting cellular function and sustaining energy availability. Over time, these adaptations increase the body's capacity to meet the ongoing physiological demands of aging.

Regulating Metabolism and Diabetes Risk

Beyond its cardiovascular benefits, exercise serves as an essential pillar of metabolic health, offering robust influences on glucose regulation, body composition, and insulin sensitivity. In sedentary individuals, insulin—the hormone that regulates blood sugar—becomes less effective, a characteristic metabolic pattern associated with impaired metabolic regulation. Regular physical movement sensitizes muscle cells to insulin, improving glucose uptake, reducing excessive insulin demand, and helping maintain stable blood sugar levels through more efficient cellular signaling.

Aerobic exercises, such as jogging or swimming, and resistance training, like weightlifting, provide distinct yet synergistic metabolic benefits. Extensive research shows that consistent physical activity is associated with lower fasting glucose levels and more favorable hemoglobin A1c profiles, a widely used marker of long-term glucose control. Moreover, exercise limits the accumulation of excess fat, especially visceral fat around vital organs, which is closely linked to chronic systemic inflammation, cardiovascular strain, and metabolic dysregulation.

By maintaining energy balance, optimizing glucose metabolism, and modulating inflammatory markers, regular movement serves as a central influence on metabolic aging. Aging need not signify inevitable metabolic decline. With every step walked, weight lifted, and gentle stretch performed, the body renews its capacity to manage energy availability, maintain metabolic flexibility, and sustain durable physical function.

Cancer Biology and Immunity

Exercise also emerges as a formidable driver of immune function and pathways involved in tumor biology. Each session of physical activity mobilizes immune sentinels—natural killer (NK) cells, macrophages, and T cells—that patrol the body, contributing to immune surveillance and cellular integrity. It is as if exercise trains a vigilant army within, primed to support the identification and clearance of abnormal cellular activity.

Research repeatedly links regular physical activity to lower observed rates of certain cancers, including breast, colon, and prostate cancers. Movement helps modulate hormone levels, diminish systemic inflammation, and enhance DNA repair mechanisms, supporting genetic integrity. Moreover, exercise influences the expression of tumor-suppressing genes, a focus of ongoing investigation in cancer biology. By boosting immune efficiency and counteracting chronic inflammation, exercise influences the biological conditions associated with tumor development.

Preserving Musculoskeletal Health and Functional Capacity

Strong muscles and healthy bones form the foundation for long-term mobility, independence, and overall well-being, and exercise offers an essential means of maintaining both. As the years pass, sarcopenia—the gradual, age-related loss of muscle mass and strength—becomes increasingly common and is associated with reduced physical capacity and increased vulnerability to instability. Picture an older adult confidently climbing stairs or gracefully carrying groceries, contrasted with someone weakened by unchecked muscle loss; the difference often lies in regular strength training. By stimulating muscle protein synthesis, preserving fast-twitch (type II) muscle fibers crucial for quick reactions, and enhancing neuromuscular coordination, strength exercises promote balance, mobility, and responsive movement.

The benefits extend beyond muscles, as bones also respond to regular physical activity. Resistance training and weight-bearing activities—such as lifting weights or brisk walking—stimulate osteoblasts, the specialized cells responsible for building and maintaining bone density. These mechanical loads strengthen the skeleton and preserve structural integrity, corresponding with lower rates of bone fragility. Everyday movements, including walking, running, or resistance exercise, not only maintain skeletal stability but also stabilize joints, enhance posture, and preserve functional movement. These effects promote physical independence and confidence, helping people remain active, capable, and self-reliant into advanced age.

Enhancing Gut Health and the Microbiome

Emerging research continues to illuminate the connection between regular exercise and gut health, underscoring how physical movement is associated with greater microbial diversity, digestive efficiency, and inflammatory regulation. The gut microbiome—an intricate community of trillions of microorganisms—plays a central role in immune function, nutrient absorption, and metabolic regulation. Studies consistently demonstrate that physical activity corresponds with increased populations of beneficial bacteria, particularly those producing short-chain fatty acids (SCFAs). Like nurturing a lush, vibrant garden within, these microbes reinforce gut-barrier integrity, regulate immune responses, and dampen systemic inflammation, highlighting exercise as an influence that extends beyond traditional metabolic pathways.

Remarkably, the relationship between the gut and exercise extends beyond digestion, engaging the brain through the gut—brain axis—a communication pathway linking digestive activity with emotional regulation, cognitive processing, and stress responsiveness. Physical activity modulates this exchange, fostering a more diverse and balanced microbiome that, in turn, is linked to measures of neural and cognitive function. Regular movement therefore affects not only digestive and immune activity, but also aligns with steadier mood and cognitive performance. Together, exercise, gut health, and brain function illustrate the deep interdependence between movement, neural vitality, and long-term physiological regulation.

Exercise and Brain Health: Cognition and Neurobiological Aging

Exercise profoundly influences brain structure and function. It modulates cognitive abilities, mental clarity, and neural adaptability—features of brain aging that intersect with biological processes studied in neurodegenerative conditions such as Alzheimer's, Parkinson's, and vascular dementia. Although aging naturally involves structural and functional brain changes—including memory decline, slower processing speeds, and diminished executive functions—regular physical activity is associated with differences in how these changes unfold. By bolstering neural integrity, exercise preserves cognitive function.

At the heart of these effects is neuroplasticity, the remarkable ability of the brain to reorganize, form new neural connections, and adapt continuously. Exercise enhances neuroplasticity by increasing levels of brain-derived neurotrophic factor (BDNF), a critical protein that supports neuron survival, strengthens synaptic connections, and fosters the formation of new neural pathways. Elevated BDNF concentrations correlate with improved memory, learning capacity, and cognitive flexibility, whereas lower levels are observed in settings of cognitive decline. Aerobic activities such as walking, running, and cycling are consistently associated with increased BDNF expression, enabling neural adaptability at all ages. Brain imaging studies further show that physically active people exhibit larger hippocampal volumes, the brain region central to memory formation and spatial navigation, highlighting structural patterns associated with regular movement.

Exercise also enhances cerebral blood flow (CBF), improving the delivery of oxygen and nutrients necessary for neuronal metabolism. With aging, vascular responsiveness often diminishes, limiting perfusion to cognitively important regions. Regular physical activity promotes angiogenesis—the formation of new blood vessels—particularly within the hippocampus and prefrontal cortex, regions involved in memory and executive processing. Improved circulation sustains neuronal metabolism and facilitates clearance of metabolic byproducts, processes implicated in oxidative stress and inflammatory signaling examined in neurodegenerative research. Physically active adults therefore often demonstrate denser cerebral vasculature, greater neural efficiency, and fewer imaging markers associated with cognitive aging, including white matter changes.

Regular exercise also engages biological processes relevant to neurodegenerative disease. In Alzheimer's disease, characterized by beta-amyloid plaques and tau pathology, physical activity influences glial cell activity and protein-handling pathways involved in neural maintenance. In Parkinson's disease, where degeneration of dopamine-producing neurons disrupts motor and cognitive function, exercise is linked to changes in dopaminergic signaling efficiency and improved motor coordination. Physical activity further modulates neuroinflammatory pathways, including inflammatory mediators such as TNF-α and IL-6, which are frequently elevated in neurodegenerative states. Improvements in insulin sensitivity and metabolic regulation associated with exercise are increasingly recognized as relevant to cognitive aging and dementia-related processes.

Different forms of physical activity engage the brain in distinct ways. Aerobic exercise is accompanied by differences in cerebral blood flow and hippocampal structure. Mind-body practices, including yoga and tai chi, influence stress-responsive neural pathways and emotional regulation. Strength training is associated with executive function and white matter characteristics. Skill-based activities such as dance, sports, and martial arts place demands on coordination, reaction time, and cognitive flexibility, reinforcing integrated brain—body communication through complex motor learning.

In this context, regular physical activity reflects sustained engagement with the neural systems that support adaptation and learning. While no single activity eliminates the possibility of neurodegenerative change, consistent movement is associated with clearer cognitive function and preserved neural adaptability. Through its effects on circulation, plasticity, metabolism, and inflammation, exercise remains a central factor in age-related brain change.

The Far-Reaching Impact of Exercise on Mental and Emotional Health

Exercise does more than benefit physical health—it influences emotional stability and psychological well-being. When the body moves, it alters brain chemistry, stimulating the release of neurotransmitters such as endorphins and serotonin. These biochemical messengers are associated with shifts in mood regulation and emotional balance. The effects of this biochemical activity can accumulate, reinforcing emotional stability, improving sleep quality, and supporting the brain's capacity to manage stress.

Chronic stress rarely announces itself in dramatic bursts. More often, its effects accumulate gradually, exerting steady pressure on both mental and physical health. Consider the constant strain it imposes—it erodes well-being like a persistent drip wearing away stone. Exercise acts as a counterweight by dampening excessive cortisol effects, the body's primary stress hormone. While short-term elevations in cortisol are essential for adaptive responses, prolonged elevation is associated with anxiety, depressive symptoms, and inflammatory signaling. Regular movement moderates this response, fostering greater calm and control. Certain practices—particularly yoga and tai chi—not only strengthen the body but also engage parasympathetic nervous system activity, the body's primary relaxation pathway. In this way, consistent physical activity is associated with an improved capacity to navigate daily stressors with clarity and composure.

Beyond biochemical balance, exercise cultivates perseverance and discipline—qualities that extend beyond the physical act itself. The challenges encountered during physical activity, whether pushing through fatigue during a run, mastering a demanding yoga pose, or completing a difficult strength workout, echo the psychological endurance required in everyday life. Strength training, endurance exercise, and high-intensity interval training engage neural circuits involved in effort regulation and motivation. Through repeated activation of these pathways, attention, self-efficacy, and emotional regulation are strengthened. Participation in structured exercise programs is associated with improved emotional regulation, lower reported anxiety, and higher self-confidence.

Exercise also fosters social connection, a crucial but often overlooked component of emotional well-being. Group workouts, team sports, or even casual walks with friends create opportunities for meaningful interaction, mutual support, and shared experience—factors associated with reduced loneliness and higher life satisfaction. People who engage in social forms of physical activity report stronger emotional well-being and greater stress tolerance than those who exercise alone. Even simple interactions—a friendly greeting during a morning jog or the camaraderie of a neighborhood yoga class—can lift mood and foster a deeper sense of community. Exercise offers more than physical fitness; it contributes to cognitive function, emotional balance, physical coordination, and the social bonds essential to psychological and social well-being.

Momentum for Life: A Lifetime of Movement and Growth

Over the course of a lifetime, movement repeatedly loads the cardiovascular system, challenges metabolism, stimulates neural circuits, and preserves muscle and bone. These effects do not occur in isolation or on a single timetable. Instead, each bout of activity leaves a small physiological imprint that accumulates across years, altering how the body repairs itself, maintains function, and tolerates stress as it ages..

Among lifestyle choices that influence lifespan and healthspan, exercise remains one of the most powerful and broadly influential behaviors available. No other habit simultaneously engages so many biological systems or exerts such wide-ranging effects on aging physiology. Genetics may provide the biological blueprint and medicine may intervene when problems arise, but movement modifies the conditions under which age-related change occurs, rather than passively responding to its effects.

Exercise is not merely protective—it functions as a broad amplifier of biological functional capacity, operating through mechanisms distinct from medical intervention. It fortifies the heart, preserves brain function, regulates metabolism, and maintains the integrity of muscle and bone—all essential for remaining physically capable and mentally engaged. Yet its influence penetrates deeper still. At the cellular level, exercise modulates aging-related processes, activating repair pathways, enhancing mitochondrial function, and sustaining the regenerative capacity of tissues and organs. More than a health practice, it is a catalyst for biological renewal.

Even small amounts of movement accumulate meaningful effects. A short daily walk, taking the stairs instead of the elevator, standing periodically throughout the day—these seemingly minor choices resonate across a lifetime, reinforcing adaptability and capability. The objective is not rigid perfection but ongoing engagement, allowing the body to remain responsive and capable throughout life.

Exercise does more than influence lifespan—it alters the biological quality and functional coherence of the years lived. Among longevity-oriented behaviors, movement endures as one of the most extensively studied and widely accessible influences on how aging unfolds.

CHAPTER 5

Nutrition for Longevity: The Science Behind Food and Health

"Let food be thy medicine and medicine be thy food."
—HIPPOCRATES

Nutrition influences the body far more deeply than meeting the basic need for sustenance; it helps determine how the body adjusts and responds as conditions change. Each meal interacts with core biological systems—metabolic pathways, cellular repair processes, and the dense microbial world of the gut. Food is not merely fuel; its molecules move through the bloodstream, engage chemical signals, and take part in the ongoing maintenance that keeps tissues functioning. These effects build gradually across days, months, and years.

Popular discussions about nutrition often fixate on calories or the latest dietary trend, yet the more important story lies in how specific foods act within the body. Nutrients regulate digestion, adjust metabolic timing, modulate inflammatory activity, and affect the chemical messengers that guide cellular and organ-level responses. Diet has been closely linked to aging-related changes, and nutrition remains one of the most direct and accessible ways to engage with the biological processes that unfold over a lifetime.

Nutrition has been approached from nearly every angle—through scientific journals, long-form books, and an endless stream of public commentary. Rather than retracing that terrain, the discussion here focuses on principles that matter in daily life. It sets aside exhaustive biochemical detail in favor of practical clarity, offering a way to think about food that aligns with the realities of modern eating.

By addressing nutrition through both scientific foundations and everyday practice, this chapter emphasizes choices that can be sustained over time. It avoids rigid rules and provides guidance that fits into ordinary routines—meals prepared at home, decisions made quickly, and habits formed over years. The aim is to translate evidence into ideas that remain clear and applicable at the moments when food decisions actually occur.

A Cellular Perspective: Nutrition, Oxidative Stress, and Inflammation

Aging unfolds at the cellular level, where trillions of cells accumulate wear from oxidative stress, metabolic strain, and low-grade inflammation. These pressures influence DNA stability, protein integrity, and how tissues respond over time. Oxidative stress—an imbalance between reactive molecules and the systems that clear them—can disrupt genetic material, distort protein structure, and weaken cell membranes. Foods containing antioxidant compounds, including blueberries, spinach, walnuts, citrus fruits, and nuts and seeds rich in vitamin E, contribute to the body's capacity to manage these oxidative demands.

Chronic inflammation is closely tied to aging biology. Diet can influence inflammatory activity, and nutrients such as omega-3 fatty acids—found in salmon, mackerel, flaxseeds, and walnuts—are linked to immune signaling. Coenzyme Q10, present in foods like spinach and organ meats and available as a supplement, has been examined for its role in mitochondrial energy production, a recurring theme in research on cellular performance.

Mitochondria generate the energy required for cellular functions, from repair to structural maintenance, yet their efficiency can decline under sustained stress. This vulnerability has made mitochondrial activity a consistent focus of aging research. Ongoing work continues to examine how nutrition relates to mitochondrial function, including whether specific nutrients influence energy output or contribute to the renewal of these organelles through processes such as mitophagy and biogenesis.

Metabolic Balance, Caloric Restriction, and Time-Restricted Eating

Metabolic stability supports many of the body's core functions as it ages. When this stability is disrupted—through excessive sugar intake, limited movement, or irregular eating—metabolic pathways can pivot toward inflammatory activity, altered insulin responsiveness, and greater oxidative stress. These shifts are increasingly **connected to**cardiovascular, neurological, and endocrine regulation.

Caloric restriction (CR), defined as reducing energy intake without compromising nutrient adequacy, has produced striking results in animal studies, including changes in lifespan across multiple species. Findings from human research show more modest effects. Even so, CR corresponds with changes in metabolic markers such as improved insulin sensitivity, steadier lipid profiles, and lower levels of inflammatory indicators. These observations guide ongoing efforts to understand how energy intake relates to metabolic function.

Strict caloric restriction can be difficult to sustain. As an alternative, time-restricted eating (TRE)—limiting food intake to a daily window such as 8 to 12 hours—has attracted interest as a more practical option. Eating within a defined daily interval corresponds with differences in metabolic regulation, including markers related to glucose control and inflammatory activity. The timing of food intake appears to influence certain aspects of metabolic behavior.

Polyphenol-rich foods such as green tea, dark chocolate, and berries have drawn scientific attention for their potential associations with cellular pathways, including autophagy—the process through which cells dismantle and recycle damaged components. Autophagy plays an important role in aging biology. Current work continues to investigate how specific nutrients relate to activity in this pathway, reflecting efforts to understand how dietary compounds intersect with the cellular machinery responsible for repair and turnover.

Caloric restriction, time-restricted eating, and nutrient-dense diets frame much of the current discussion about how energy intake and meal timing relate to metabolic function during aging. Their relationships to glucose control, inflammatory activity, and other biological measures help clarify why these practices continue to draw attention. Chapter 7 examines these dietary choices in greater depth, outlining the mechanisms involved and the practical considerations that arise when they are applied in everyday life.

The Gut Microbiome: Nutrition's Silent Partner

Hidden within the digestive tract lies a dense community of microorganisms that exerts measurable influence on digestion, immune activity, and a range of other physiological functions. Known as the gut microbiome, this ecosystem communicates with the body through chemical signals, microbial metabolites, and direct contact with the intestinal lining. Dietary fiber plays a central role in sustaining this community by providing substrate for bacteria that generate short-chain fatty acids—molecules involved in maintaining barrier integrity and modulating immune responses. Fiber-rich foods such as beans, oats, leafy greens, and whole grains supply prebiotics that support the expansion of these organisms and help maintain microbial diversity.

Probiotics offer another way of engaging with the gut microbiome. These live microorganisms, found in fermented foods such as yogurt, kefir, kimchi, sauerkraut, and miso, can introduce strains that interact with existing bacterial communities and contribute to metabolic activity within the intestinal tract. Probiotic intake has been examined in relation to digestive function, inflammatory activity, and neural signaling within the gut—brain axis—the bidirectional system linking the intestine with the central nervous system.

Certain eating habits can shift the gut microbiome in less favorable directions. High intake of ultra-processed foods, refined carbohydrates, or artificial sweeteners has been linked in population studies to reduced microbial diversity and increases in bacterial groups connected to metabolic dysregulation. These microbial shifts correspond with inflammatory activity, altered insulin responsiveness, and other elements of metabolic regulation. Choosing whole foods and fiber-rich ingredients supports a more resilient microbial environment.

The Mediterranean Diet: A Model for Longevity?

Among dietary patterns examined in aging research, the Mediterranean diet stands out for the breadth and consistency of its findings. Rooted in long-standing food traditions of Mediterranean regions, it centers on whole, minimally processed ingredients such as fruits, vegetables, whole grains, nuts, seeds, and extra virgin olive oil. Moderate portions of fish, poultry, and dairy products supply additional protein sources, while red meat and processed foods appear far less often. Rather than emphasizing restriction, this way of eating relies on straightforward, nutrient-dense foods that have supported daily life in these regions for generations.

Cardiovascular findings anchor much of the early work. In the PREDIMED trial, Mediterranean-style eating supplemented with extra virgin olive oil or nuts was associated with lower rates of major cardiovascular events compared with a low-fat diet. A 2019 meta-analysis reported similar results, noting that people with greater alignment to this eating pattern tended to show lower cardiovascular mortality. These associations are often considered alongside measures of inflammation, lipid handling, and vascular responsiveness.

Observations in aging populations reinforce this picture. Work in regions often described as "Blue Zones"—areas noted for a higher proportion of older adults who remain active and cognitively capable—frequently highlights eating habits that resemble Mediterranean traditions. In Sardinia, for example, daily meals commonly include legumes, whole grains, and olive oil, a practice that has persisted for generations and remains woven into everyday life.

Mortality findings offer a broader view. A 2018 meta-analysis published in the *British Journal of Nutrition* evaluated how consistency with the Mediterranean diet related to all-cause mortality. Across diverse cohorts, stronger adherence corresponded to lower mortality, underscoring its role as a central point of comparison in long-term dietary research.

Research on cognitive aging presents a similar picture. A study published in *Neurology* reported that stronger adherence was linked to slower cognitive decline and a lower likelihood of Alzheimer's-related features over the study period. The diet's concentration of polyphenols, antioxidants, and omega-3 fatty acids has been examined in relation to neuroinflammatory activity, oxidative processes, and signaling pathways involved in neuronal support. The MIND diet—a hybrid of the Mediterranean and DASH diets—has likewise been evaluated for cognitive outcomes, with several analyses finding more favorable performance on measures of memory and executive function among those who followed it closely.

Metabolic findings show comparable trends. A study published in *Diabetes Care* reported that people who ate in this style showed improvements in insulin sensitivity and steadier glucose control. Additional reviews have identified relationships between Mediterranean-style eating and weight stability over time. Together, these findings indicate a consistent relationship between this way of eating and markers of metabolic regulation.

Cancer-related outcomes form another area of investigation. A 2020 review published in *Nutrition Reviews* reported that higher adherence to the Mediterranean diet corresponded to lower incidence of colorectal, breast, and gastric cancers across several cohorts. These findings are commonly interpreted in light of the diet's concentration of antioxidants, polyphenols, and other constituents that influence inflammatory and oxidative processes.

What distinguishes the Mediterranean diet is the combined influence of its core components. Healthy fats, plant-derived antioxidants, and fiber each interact with pathways tied to inflammation, oxidative stress, and metabolic regulation. Olive oil, nuts, and fatty fish supply lipids that affect membrane composition and cellular signaling. Fiber supports the gut microbiome and fuels the production of short-chain fatty acids, which contribute to intestinal and immune function. These components outline a way of eating consistently that aligns with steadier metabolic function and sustained physiological capacity.

The Power of Energy Balance: Understanding Calories In vs. Calories Out

Diets such as keto, Mediterranean, or intermittent fasting often dominate nutrition conversations, yet they all hinge on a shared principle. Energy balance—the relationship between calories consumed and calories expended—remains central to weight regulation and to the metabolic and hormonal responses that accompany changes in intake. When energy intake exceeds expenditure, the body stores the excess; when expenditure surpasses intake, stored reserves are used.

This principle affects not only body weight but also how metabolic and hormonal systems operate. It influences metabolic pathways and hormone activity, including how the body handles fuel, absorbs and distributes nutrients, and responds to meals throughout the day. Maintaining a body weight that can be supported efficiently often corresponds with changes in insulin responsiveness, inflammatory activity, and more stable metabolic regulation. While food quality remains important, understanding how energy intake tracks with expenditure provides a clearer basis for interpreting long-term changes in weight.

Energy balance also allows for different eating styles. A plant-based pattern, a low-carbohydrate plan, or a mixed approach shaped by taste and routine can all operate within this framework, as energy balance continues to govern how the body uses and stores energy. Rather than following dietary trends, recognizing how energy intake relates to expenditure helps people make choices that remain workable.

Nourishing a Long and Healthy Life

Nutrition affects far more than the basic need for sustenance. It influences how cells produce energy, how metabolic pathways respond to meals, and how immune activity varies throughout the day. Each food choice interacts with systems involved in energy production, inflammatory responses, and cellular maintenance—mechanisms that guide how the body repairs and adapts. These effects accumulate and contribute to the biological experience of aging.

Specific foods participate in physiological processes relevant to aging. Antioxidant-rich fruits such as blueberries and leafy greens like kale have been studied for their associations with oxidative-stress responses. Whole grains—including quinoa, oats, and barley—support digestive function and contribute to steadier energy use. Fats from avocados and olive oil appear in studies focused on cardiovascular measures and membrane composition. Nuts and seeds supply micronutrients involved in structural maintenance and metabolic activity. Considered together, these foods reflect broader dietary habits linked to metabolic, cognitive, and emotional outcomes.

Food practices carry cultural meaning as well as biological effects. They reflect heritage, identity, and shared experience. Family recipes and meals prepared with care can reinforce connection and strengthen community ties. This raises a practical question: how can nutritional insight fit alongside longstanding culinary traditions? One response is to choose ingredients that respect cultural preference while supporting physiological priorities. A family gathering can include traditional dishes prepared with thoughtful choices that preserve meaning while avoiding unnecessary excess.

In the end, nutrition succeeds not through discipline alone, but through continuity. The most durable choices are those that respect appetite, culture, and circumstance while remaining grounded in biological need. When eating aligns with both physiology and lived experience, it becomes less an intervention and more a sustaining practice—one that carries forward.

CHAPTER 6

Supplements and Functional Foods: Precision Nutrition and Aging

"Knowledge is the food of the soul."
—PLATO

Dietary supplements and functional foods occupy a distinct and often misunderstood place within the broader landscape of longevity. They are neither miracle agents nor trivial additions. They are better understood as targeted tools that can support a well-constructed diet and lifestyle. Nourishing food remains the foundation on which healthy aging rests, yet certain supplements may offer added reinforcement as the body adjusts to changes that accompany time: shifts in inflammatory activity, the gradual burden of oxidative stress, alterations in mitochondrial function, and the growing influence of senescent cells. Autophagy—the essential process that removes damaged cellular components—also tends to slow with age, creating conditions in which thoughtful nutritional choices may be relevant.

Even so, supplements do not act uniformly across all people. Genetics, diet, daily habits, and current health status impact how any compound is absorbed, metabolized, and utilized. They do not replace a balanced base; they work alongside it. A useful way to think about this is through the image of a well-built home. Nutritious food, regular physical activity, restorative sleep, and effective stress management create its structural integrity. Supplements can then serve as finishing elements—refinements that strengthen the whole design, but only when the underlying structure is sound.

This chapter examines supplements and functional foods with stronger evidence for effects on biological processes relevant to aging, including metabolism, cellular maintenance, and physiological capacity.

The Practical Role of Supplements in Aging

Supplements can contribute by offering targeted support—addressing nutritional gaps or reinforcing biological processes that change with age. Omega-3 fatty acids provide a clear example. Evidence across multiple study designs points to their involvement in inflammatory regulation and neurological function, with higher intake aligning with steadier cognitive performance and measures related to brain activity. Vitamin D offers another illustration. It participates in calcium balance and musculoskeletal integrity, and lower levels are common in regions with limited sunlight. Research also consistently notes links between insufficient vitamin D and reduced bone mineral density.

Even with these findings, expectations should remain realistic. Supplements seldom produce rapid or dramatic changes. Their influence builds gradually and is most effective when combined with the advantages created by whole foods. Antioxidant-rich foods—berries, leafy greens, nuts—contain nutrients arranged in ways that isolated compounds rarely match. Still, targeted supplementation can sometimes reinforce existing cellular responses initiated by these foods. Pairing a polyphenol-rich diet with certain antioxidants, for example, has been examined in relation to cellular responses involved in managing oxidative stress. Yet isolated compounds have clear limits: effects observed in controlled laboratory settings are often smaller in human studies. Current investigations seek to clarify why whole-food compositions can deliver outcomes that single nutrients do not consistently reproduce.

Seen in a broader context, supplements provide their greatest value when incorporated into a balanced way of living supported by nutritious food, regular physical activity, adequate sleep, and steady stress management. Within that foundation, supplements serve as practical tools—enhancing mental clarity, physical capability, and day-to-day functional stability as the body moves through the aging process.

Omega-3s and Vitamin D: Foundations for Brain and Bone Support

Omega-3 fatty acids—particularly EPA and DHA, found in salmon and other cold-water fish—play roles that extend beyond cardiovascular physiology. Research has examined how these fats contribute to neuronal membrane structure, influence membrane fluidity, and support communication between brain cells. Higher intake corresponds with clearer thinking and steadier memory performance in several observational studies, and some long-term analyses describe cognitive trajectories that show greater stability than those observed in people with consistently low intake. Early work in neurodegenerative conditions has reported findings of interest as well, though results vary and the biological mechanisms are still being investigated. For those who eat little fish, plant sources such as flaxseeds and walnuts provide ALA, a precursor that the body converts to EPA and DHA only in limited amounts. Supplements have therefore been explored as a practical means of increasing EPA and DHA intake without substantial dietary change.

Vitamin D helps regulate calcium absorption and balance, most notably within the skeletal system. Many adults—especially those living at northern latitudes or spending limited time outdoors—show low circulating levels. Studies across diverse populations have reported associations between lower vitamin D status and reduced bone mineral density. Some investigations also examine the combined use of vitamin D3 and vitamin K2, observing that the pairing may influence how calcium is distributed within the body, though findings remain mixed and continue to evolve as additional evidence emerges. While no definitive intake standard for vitamin K2 in this pairing has been established, the accumulated research helps explain why these nutrients are often discussed together when considering long-term skeletal support.

Creatine: An Athletic Supplement Reimagined for Aging

Creatine is often linked with strength training and athletic performance, yet its relevance reaches far beyond the gym. As muscle tissue changes with age—losing power, recovering more slowly, and gradually declining in mass—creatine has been studied for its role in the body's short-term energy reserve. This reserve can be mobilized quickly, helping muscles manage brief bursts of effort and repeated movements. Many controlled trials report that creatine use is associated with gains in muscle performance among older adults. These findings do not guarantee protection from injury or falls, but they consistently show that people with greater muscular strength often move with more confidence and stability in everyday life.

Interest has also grown in creatine's influence on cognition. Studies involving older adults or people experiencing fatigue-related cognitive slowing have reported associations with improvements in memory tasks, reaction speed, and general mental clarity. These effects are modest rather than dramatic, yet their reproducibility in certain groups has sustained scientific attention. With a long research record, low cost, and straightforward use, creatine occupies a practical place within nutritional strategies examined for supporting physical and cognitive function as the body ages.

CoQ10 and NAD+: Cellular Energy and the Architecture of Vitality

Many people describe aging as a gradual loss of internal energy—a sense that the body's "battery" no longer holds its charge as well as it once did. On a biological level, this perception aligns with changes in mitochondrial function. These organelles, responsible for producing the cell's metabolic output, tend to work less efficiently over time. Coenzyme Q10 (CoQ10) is central to this energy-transfer process, and studies show that its levels often decline with age in several tissues. Some medications, including statins, have been reported to influence CoQ10 availability, prompting interest in whether supplementation has been studied for its role in sustaining mitochondrial performance. Several controlled trials have reported improvements in markers related to cellular energy use and cardiovascular function in association with higher CoQ10 levels, although the strength of these effects varies among studies.

NAD+ has gained attention for its involvement in metabolic activity, DNA repair, and broader cellular maintenance. Levels of this molecule also drop with age, raising questions about whether increasing NAD+ through precursors such as NMN or NR relates to metabolic health. Early human studies have noted shifts in mitochondrial activity and related metabolic indicators, but the field remains under active investigation. Important questions remain about long-term safety, dosing strategies, and how these findings translate into day-to-day outcomes. Because NAD+ plays a prominent role in discussions of aging biology, a later chapter examines NMN, NR, and NAD+ in greater depth and places current research within the larger context of cellular aging.

Antioxidants and Aging: Resveratrol, Polyphenols, and Curcumin

Resveratrol, found in red wine, grapes, and berries, has long drawn interest because of its effects in laboratory models of aging. Dietary intake does not raise resveratrol levels in a meaningful way—the amounts present in food are too low. Supplements, by contrast, provide higher concentrations that have been used to study resveratrol's interactions with sirtuins, a family of proteins involved in stress responses, inflammation, and DNA repair.

Findings from cell and animal work can be striking, but translating them into clear human outcomes has proved difficult. Some human studies report changes in metabolic or inflammatory markers, while others show more limited effects. These differences underscore the complexity of the pathways involved and help explain why continued research is needed.

Polyphenols, abundant in fruits, vegetables, olive oil, dark chocolate, and tea, form a broad family of compounds with wide-ranging biological activity. Observational studies often link higher polyphenol intake with cardiovascular and cognitive advantages across the lifespan. These compounds influence inflammatory signals and oxidative stress in gradual, cumulative ways, contributing to a more stable physiological environment.

Curcumin, the primary active compound in turmeric, continues to attract close scientific attention. Studies examining its associations with inflammation, oxidative stress, and joint comfort report varied results, including findings of interest in groups with arthritis-related symptoms. Many trials document improvements in self-reported comfort or mobility. Curcumin's low bioavailability, however, remains a practical constraint. To address this, supplements commonly combine curcumin with piperine (black pepper extract), which enhances absorption. Even with enhanced formulations, outcomes differ across studies, and research continues to determine how curcumin is best applied.

Clearing Senescent Cells: Fisetin and Quercetin

Senescent cells resemble biological machinery that no longer functions as intended yet refuses to shut down. Instead of completing their life cycle and exiting, they persist and release inflammatory signals that can influence nearby tissues. Their gradual buildup is a well-recognized feature of aging, and interest has grown in compounds that have been studied for their potential role in helping the body manage this burden.

Fisetin, found in strawberries, and quercetin, present in apples, onions, and capers, are flavonoids that have shown notable effects in laboratory studies. In cell cultures and animal models, both compounds have reduced senescent-cell burden and have been associated with improvements in tissue function and inflammatory measures. These senolytic effects, while striking in controlled settings, are still being evaluated in humans. Early trials and pilot studies have reported measurable biological effects, but the evidence is preliminary. Important questions about dosing, long-term safety, and how well laboratory findings translate to complex human biology are under active study. Even so, the underlying concept retains interest: supporting the body's endogenous processes for clearing damaged cells aligns with broader efforts to understand cellular aging.

Cellular Housekeeping with Spermidine

Autophagy is one of the body's key maintenance processes. It identifies damaged cellular components, breaks them down, and recycles what can still be used. With age, this system tends to work less efficiently, allowing damaged material to accumulate. Spermidine, a naturally occurring polyamine found in soybeans, wheat germ, mushrooms, legumes, and certain aged cheeses, has gained attention because it has been studied for its capacity to stimulate autophagic activity in experimental settings.

Observational research has reported that people who consume more spermidine-rich foods often show more favorable cardiovascular markers and, in some cohorts, longer average lifespans. Early human studies have examined possible effects on cognition, inflammatory activity, and day-to-day functioning. Some results are promising, though the evidence is early and incomplete. Even so, spermidine's presence in common foods and its long culinary history make it a practical element of dietary choices relevant to biological processes that weaken with age.

Gut Health: Probiotics, Prebiotics, and Fermented Foods in Longevity

The gut microbiome functions as a densely interconnected ecosystem, influencing far more than digestion. It plays a part in immune activity, metabolic regulation, inflammatory pathways, and aspects of cognitive and emotional health. With age, microbial diversity often declines, and this shift has been associated with increases in inflammation and reduced physiological flexibility. Fermented foods—such as yogurt, kefir, kimchi, miso, and sauerkraut—introduce live microbes that have been studied for their relationship to microbial diversity. Human studies have noted improvements in digestive comfort, certain immune markers, and inflammatory measures among people who regularly consume fermented foods, though outcomes vary across populations.

Probiotics, whether obtained through food or supplements, have been examined for their associations with changes in gut composition. Many studies suggest that probiotics can encourage the growth of beneficial bacterial groups and enhance gut barrier function and metabolic balance. These microbes rely on adequate nourishment to take hold. Prebiotics—fibers found in foods like onions, garlic, bananas, oats, and whole grains—provide this nourishment. Together, prebiotics and probiotics help create conditions that **foster** a stable and diverse microbial community, contributing to biological environments linked to healthier aging.

Fiber Supplements: Supporting Gut and Metabolic Health

Ideally, fiber comes from whole foods—vegetables, fruits, legumes, seeds, and whole grains. Yet aging can influence appetite, digestion, and eating routines, making it difficult for many people to meet recommended intake through diet alone. Fiber supplements such as psyllium husk and inulin have been studied for their roles in lipid metabolism, glucose regulation, and digestive function. Controlled trials often report that psyllium intake is associated with lower LDL cholesterol and steadier blood glucose levels. These findings reflect quantifiable changes in metabolic biomarkers.

Inulin, another widely examined fiber, acts primarily as a prebiotic by selectively nourishing beneficial gut bacteria. Studies have reported that adding inulin to the diet is associated with changes in microbial composition linked to improved metabolic markers. Supplements are not a substitute for fiber-rich foods, but they serve as practical tools when dietary intake falls short, helping maintain a digestive environment that supports metabolic regulation.

Collagen Peptides and Magnesium: Structural Support Across the Lifespan

Collagen supports the structure of skin, bones, joints, and connective tissues. Over time, natural collagen production declines, contributing to visible and functional changes such as reduced elasticity, joint stiffness, and shifts in tissue strength. Research on collagen peptides has grown quickly, and many trials report associations with improvements in skin hydration, elasticity, and self-reported joint comfort. These outcomes vary across individuals, but the consistency of results suggests that collagen peptides are being examined for their role in reinforcing structural tissues in ways that align with the body's changing biology.

Magnesium is involved in hundreds of physiological reactions, including energy production, glucose regulation, muscle contraction, nerve signaling, and bone maintenance. Inadequate intake is relatively common, particularly in older adults. Low magnesium levels have been associated with cardiovascular, metabolic, and musculoskeletal concerns. Food sources—leafy greens, legumes, nuts, and seeds—remain the preferred foundation, but supplementation has been studied as an option when dietary intake is insufficient. Trials have reported associations with improvements in sleep quality, insulin sensitivity, and muscle performance, though findings depend on population and context. Magnesium contributes to biochemical processes that fit naturally within whole-food nutrition.

Green Tea and EGCG: Everyday Support for Metabolism and Cellular Health

Green tea has held cultural significance for centuries, and scientific interest in its bioactive compounds—especially EGCG—continues to grow. EGCG has been studied for its antioxidant activity and its involvement in pathways related to inflammation, metabolism, and cellular protection. Studies of populations with high green tea consumption often report associations with favorable cardiovascular markers and steadier cognitive performance. Controlled trials have also reported associations with modest improvements in blood sugar control, cholesterol measures, and other metabolic indicators among people who consume green tea or moderate amounts of EGCG extracts.

The connection between dose and effect requires careful consideration. Concentrated, high-dose extracts can strain the liver in susceptible individuals, leading researchers to emphasize moderation. Drinking green tea provides a more balanced intake of EGCG and other polyphenols within the nutritional environment in which they naturally occur. For those who use extracts, research settings tend to favor moderate amounts, reflecting a broader view that nutritional compounds are most helpful when they work alongside the body's regulatory systems rather than overwhelming them.

Whole Grains, Nuts, and Seeds: Building Blocks for Metabolic Stability

Whole grains such as oats, barley, buckwheat, and quinoa provide complex carbohydrates that release energy gradually, producing fewer sharp glucose swings than those seen with refined grains. Their fiber content supports beneficial gut microbes, and many studies associate whole-grain intake with improvements in digestive and metabolic markers. Even small substitutions—like choosing whole-grain bread over refined alternatives—can, over time, contribute to steadier metabolic regulation in ways that accumulate.

Nuts and seeds offer concentrated nutrition. Almonds, walnuts, flaxseeds, chia seeds, and pumpkin seeds supply fiber, antioxidants, vitamins, minerals, and omega-3 fatty acids. Research often links regular nut consumption with healthier cardiovascular markers, reduced inflammatory activity, and more stable cognitive performance in certain groups. Their influence is not immediate, but the consistency of findings across large population studies illustrates a common theme in longevity research: modest habits practiced over many years tend to have the greatest impact.

Multivitamins: Filling Nutritional Gaps

Even well-planned diets can leave nutritional gaps, especially as aging affects appetite, digestion, and nutrient absorption. Vitamins such as B12, folate, magnesium, and zinc may become harder to obtain in adequate amounts, and low levels are often discussed in relation to cognitive, cardiovascular, and metabolic processes. Several large clinical trials have evaluated multivitamin use in older adults, with some reporting associations with improvements in cognitive test performance, memory measures, or nutritional status. Results differ among studies, but a consistent theme in a number of them is that multivitamins can act as a practical safety net—helping maintain access to essential nutrients when dietary variety is limited.

A multivitamin cannot reproduce the complexity of whole foods, nor is it meant to. Its role is to provide dependable baseline coverage, strengthening the nutritional foundation on which more specific supplements and food choices can build.

A Thoughtful Approach to Nutrition and Longevity

Supplements and functional foods are not shortcuts, nor do they override the interplay of biology, daily behaviors, and environmental influences that shape the aging process. They function more like careful instruments within a larger ensemble—supporting essential physiological processes while depending on the steady rhythm set by foundational habits. Compounds such as creatine, spermidine, polyphenols, probiotics, and green tea extracts do not reverse aging, but they engage cellular pathways involved in stress responses, mitochondrial activity, inflammatory signaling, and autophagy. Their role is not dramatic change but steady reinforcement for systems already working to maintain balance.

Precision remains vital. Excessive supplementation can disrupt the same processes it aims to assist. High-dose antioxidant supplements, for example, may dampen stress signals needed for healthy cellular adaptation. As fields such as genomics, wearable monitoring, biomarker analysis, and nutrition science advance, supplementation strategies may become more tailored, reflecting a deeper understanding of how physiology varies across people and across time.

Viewed more broadly, health across the lifespan grows from habits that anchor the body—nutrient-rich foods, regular physical activity, restorative sleep, and effective stress management. Supplements contribute by filling nutritional gaps, engaging key pathways, and supporting functions that change with age. Used with care, they can help sustain mental clarity and physical capability, enriching both longevity and the day-to-day experience of growing older.

CHAPTER 7

Fasting and Caloric Restriction: What Happens Inside the Body

"To lengthen thy life, lessen thy meals."
—BENJAMIN FRANKLIN

Could the path to healthier aging lie not only in the foods we consume, but also in when and how much we eat? Throughout most of human history, fasting was not a deliberate wellness practice—it was an unavoidable reality created by fluctuating food availability. These alternating periods of feast and scarcity shaped human physiology, favoring traits that supported survival during limited intake and helped maintain function when energy was scarce. In contrast, modern abundance has largely eliminated these natural intervals, a shift that has been linked to physiological strain and greater vulnerability to age-related metabolic issues. Deliberately revisiting elements of these ancestral rhythms—through approaches such as caloric restriction and intermittent fasting—has been linked to changes in insulin signaling, cellular stress—response pathways, and other biological markers involved in cellular maintenance. These strategies offer a structured way to explore how the body responds to varying periods of energy intake.

Far from simple weight-loss trends, these nutritional cycles engage deeply conserved biological mechanisms that have guided human physiology for millennia. They influence energy metabolism, shift hormonal signaling, reduce inflammatory signaling, and activate the body's intrinsic systems for cellular maintenance. Unlike conventional diets that emphasize short-term outcomes, caloric restriction and intermittent fasting operate at a molecular level, interacting with processes connected to aging biology and long-term physiological regulation.

Caloric Restriction: A Controlled Approach to Longevity

Caloric restriction (CR) represents much more than reducing food intake—it is a carefully calibrated practice of lowering energy consumption while preserving essential nutrition. For nearly a century, scientists have investigated CR's biological effects, particularly in animal models. In a wide range of species, reducing caloric intake by 20 to 40 percent has consistently extended lifespan and delayed the onset of age-related changes in laboratory settings. In humans, the question of lifespan extension remains under active study, but research such as the CALERIE trial (Comprehensive Assessment of Long-term Effects of Reducing Intake of Energy)—a rigorous, controlled investigation of moderate caloric restriction in healthy adults—has reported improvements in markers related to inflammation, cardiovascular function, insulin sensitivity, and metabolic regulation.

At the cellular level, caloric restriction prompts a shift toward greater metabolic efficiency. The body responds to this nutritional stress by conserving energy, repairing accumulated cellular damage, and reallocating resources toward maintenance rather than growth. A central component of this adaptive response is autophagy, the body's intrinsic recycling system that clears damaged or dysfunctional cellular components before they accumulate. Through this process, cells reduce the burden of impaired structures that would otherwise compromise cellular performance. Caloric restriction is also linked with improved mitochondrial dynamics, enabling cells to manage oxidative byproducts more effectively—one of the contributors to cellular decline. The result is a more balanced cellular environment better equipped to support long-term function.

Beyond its cellular influence, CR affects several hormonal pathways. It modulates insulin-like growth factor-1 (IGF-1), a protein tied to growth and cellular proliferation but also relevant to aging biology. In animal models, lower IGF-1 levels are associated with extended lifespan, reflecting a shift from growth toward preservation. In humans, this relationship appears more nuanced. Moderate reductions in IGF-1 may correspond with healthier aging markers, whereas excessively low levels in older adults have been linked with frailty and decreased muscle mass—illustrating the need for balance when considering these pathways.

Another important component of CR's biological influence involves sirtuins, a family of proteins often described as key coordinators of cellular adaptation, DNA repair, and mitochondrial function. Caloric restriction activates sirtuins, enhancing the body's capacity to respond to metabolic stress and maintain cellular integrity. Together, these effects suggest that CR is not merely a dietary adjustment but a broad biological shift that engages systems central to aging physiology.

Through this interplay of cellular maintenance, hormonal recalibration, and metabolic adaptation, caloric restriction emerges as a compelling model for understanding how nutritional strategies may support healthspan and moderate aspects of physiological decline. The improvements observed in studies such as CALERIE highlight CR's relevance to processes linked with aging. Caloric restriction demonstrates how strategic nutritional choices can influence biological rhythms and support physical and cognitive function in ways that align with long-term well-being.

Intermittent Fasting: The Power of Meal Timing

Intermittent fasting (IF) is less about restricting calories and more about structuring when meals are consumed. Instead of focusing solely on the composition of food, IF emphasizes the timing of nourishment, aligning with the body's long-standing ability to shift between sources of energy. Unlike continuous caloric restriction, intermittent fasting alternates between periods of eating and fasting, prompting distinct physiological responses during each phase.

Intermittent fasting encompasses several structured eating patterns. In scientific and clinical contexts, time-restricted eating typically involves limiting food intake to a defined window within a 24-hour period. Alternate-day fasting and reduced-intake "fasting days" introduce intermittent periods of lower energy intake, while extended fasts lasting more than 24 hours appear in some research settings, usually under supervision. Together, these formats illustrate the range of methods used to study metabolic adaptation to intermittent energy reduction.

During fasting intervals, the body undergoes a gradual metabolic shift. As glycogen stores in the liver and muscles diminish, metabolism transitions from glucose toward fat oxidation, producing ketones as an alternative fuel. Ketones serve not only as an efficient energy source—particularly for the brain—but also as signaling molecules that influence oxidative stress and cellular protection. Across multiple studies, fasting has been associated with improved insulin sensitivity, steadier glucose regulation, and shifts in metabolic markers that reflect the ability to adapt to changing energy availability.

Intermittent fasting, like caloric restriction, activates autophagy—the body's internal housekeeping machinery. When energy intake decreases, cells initiate processes that dismantle accumulated proteins, malfunctioning organelles, and other forms of molecular debris. This response supports cellular maintenance under conditions of stress. Fasting has also been examined for its effects on brain-derived neurotrophic factor (BDNF), a protein involved in neuronal growth and aspects of cognitive function. Some studies report increases in BDNF levels during fasting cycles, suggesting a link between intermittent energy restriction and brain physiology.

Inflammatory markers frequently shift in IF studies as well. Numerous controlled trials have reported reductions in markers linked to inflammation and changes in lipid profiles and blood pressure. While results vary by physiology, fasting duration, and dietary context, cumulative research suggests that intermittent fasting engages biological pathways relevant to metabolic and cellular stability.

Structured fasting periods offer a way to examine the body's capacity for metabolic adaptation and cellular maintenance. More than a dietary technique, intermittent fasting reflects feeding—fasting cycles that were common across much of human history, underscoring that the timing of nourishment carries biological significance alongside its composition.

How CR and IF Influence Longevity Biology

Both caloric restriction (CR) and intermittent fasting (IF) activate deeply conserved pathways involved in metabolic efficiency, cellular maintenance, and the body's capacity to adapt to stress. Rather than maintaining a continuous state of energetic abundance, these strategies introduce periodic nutritional challenge—a measured tension that alters how the body allocates and utilizes resources. This adaptive recalibration is not a modern invention but a return to feeding—fasting cycles that influenced human physiology across evolutionary time.

One notable adaptation involves the gradual shift toward ketogenesis during extended fasting periods. As glycogen stores diminish, the body begins producing ketones—molecules that function not only as an alternative fuel but also as signaling agents that influence oxidative stress, inflammatory activity, and cellular defense pathways. Ketones have been shown to activate regulators such as Nrf2, which enhances antioxidant responses, while moderating complexes like the NLRP3 inflammasome, often examined for their role in inflammatory signaling. The influence of caloric restriction and intermittent fasting extends beyond ketone metabolism. Even without sustained ketosis, these nutritional strategies engage cellular repair processes, modify mitochondrial function, and alter energy utilization under conditions of reduced nutrient availability.

At the genetic level, caloric restriction and intermittent fasting modulate the activity of proteins involved in cellular maintenance. Sirtuins—often discussed in the context of aging biology—coordinate processes such as DNA repair, mitochondrial regulation, and cellular stress responses. AMPK, the body's central energy sensor, shifts metabolism away from storage and toward utilization, promoting fat oxidation and more efficient glucose handling. Insulin-like growth factor-1 (IGF-1), a hormone involved in growth and aging biology, presents a more complex profile: animal models consistently link lower IGF-1 signaling to extended lifespan, while human data indicate that both elevated and suppressed levels may carry risk. Caloric restriction and intermittent fasting can modestly reduce IGF-1, particularly when protein intake is lower, moving signaling toward ranges associated with cellular repair while preserving essential physiological function.

Another central pathway affected by CR and IF is mTOR (mechanistic target of rapamycin), a regulatory hub governing cell growth, protein synthesis, and aspects of aging biology. When nutrients are abundant, mTOR favors growth and proliferation; when nutrients are scarce, mTOR activity declines, allowing autophagy and other repair mechanisms to take precedence. CR and IF naturally modulate this pathway, enabling the body to alternate between periods of growth and renewal. This oscillation reflects a biological rhythm observed in many long-lived species and highlights the importance of balance: excessive suppression of mTOR can impair immune function or reduce muscle maintenance, while intermittent modulation supports repair without undermining core physiological strength.

By engaging these interwoven pathways, caloric restriction and intermittent fasting influence cellular maintenance and metabolic regulation. Controlled studies show improvements in metabolic markers, inflammatory activity, and stress-response pathways—effects that reflect measurable changes in biological processes relevant to aging.

Current Research on CR and IF: What We Know So Far

Animal research has provided detailed insight into the biological effects of caloric restriction (CR) and intermittent fasting (IF). In rodents, sustained caloric restriction frequently extends lifespan while delaying age-related changes such as cancer, metabolic disruption, and neurodegeneration. In primates, CR has repeatedly improved markers of cardiovascular function and inflammation, although effects on lifespan have varied across studies. Intermittent fasting, while less consistently linked to lifespan extension in animals, produces wide-ranging physiological effects across species, including changes in metabolic adaptability, mitochondrial function, and cellular maintenance. These findings indicate that intermittent reductions in energy intake activate conserved biological responses involved in adaptation to energetic stress.

Human studies, by contrast, cannot directly measure lifespan within practical research timeframes, but they provide evidence relevant to healthspan—the portion of life marked by preserved functional and metabolic capacity. The CALERIE trial offers one of the clearest examinations of moderate caloric restriction in healthy adults. Participants who reduced calorie intake modestly over two years showed improvements in markers of inflammation, cardiovascular physiology, insulin sensitivity, and metabolic regulation. They document measurable changes in biological markers linked to aging.

Human studies of intermittent fasting reflect similar patterns. Across clinical trials, intermittent fasting produces changes in insulin sensitivity, lipid profiles, blood pressure, and other measures relevant to metabolic regulation. Some investigations of cognitive outcomes report modest improvements in areas such as memory and executive function, particularly among people with underlying metabolic dysregulation. Although results vary by fasting method, study duration, and individual physiology, intermittent fasting consistently engages metabolic adaptability—the capacity to transition efficiently between glucose and fat metabolism as energy demands change.

Metabolic adaptability supports cellular maintenance by limiting oxidative stress, improving energy utilization, and maintaining steadier metabolic balance. Although the long-term implications of caloric restriction and intermittent fasting for human longevity remain uncertain, controlled studies show consistent engagement of pathways involved in metabolic regulation and cellular maintenance as aging progresses.

Beyond Caloric Restriction: Toward Metabolic Precision

As aging research has advanced, attention has turned to strategies that seek to reproduce selected biological effects of caloric restriction (CR) and intermittent fasting (IF) without requiring sustained reductions in energy intake. One such strategy is the fasting-mimicking diet (FMD), which is designed to induce metabolic features of fasting while providing limited, structured nutrition.

In research settings, FMD typically involves consuming defined nutrient compositions over several consecutive days. During these cycles, metabolic markers shift in ways consistent with a fasting state, including reductions in insulin levels, changes in fuel utilization, and activation of autophagy. FMD can engage fasting-related pathways intermittently, without the prolonged energy restriction required by continuous fasting protocols.

This line of inquiry reflects a broader move toward individualized strategies in longevity research. Evidence increasingly shows substantial variation in how people respond to CR and IF, influenced by factors such as genetics, metabolic history, hormonal regulation, sleep patterns, and gut microbiome composition. Under identical fasting conditions, some individuals show pronounced metabolic changes, while others exhibit more modest responses, underscoring differences in baseline physiology rather than failure of the intervention.

Such variability supports a shift away from uniform protocols toward calibrated cycles of nutritional stress and recovery, guided by physiological context and biomarker data. Rather than continuous restriction, emerging models emphasize periodic engagement of adaptive pathways followed by intervals of maintenance.

This perspective also reflects a reassessment of long-standing assumptions in aging research. Continuous caloric abundance does not appear to maximize engagement of cellular repair mechanisms. Short, intermittent reductions in energy availability activate pathways that remain less engaged under constant feeding conditions. These cycles resemble environmental conditions under which human metabolic systems evolved, marked by alternating periods of availability and scarcity. In this framework, structured fasting strategies offer a controlled way to examine how modern physiology responds to intermittent metabolic challenge and recovery.

Integrating Precision Strategies Into Everyday Life

Tailoring caloric restriction and intermittent fasting to individual physiology reflects a move away from rigid rules and fixed calorie targets. Instead, these strategies engage cellular repair, metabolic regulation, and stress-response pathways in ways that match the body's capacity to adapt. Even modest changes—such as adjusting meal timing or incorporating periodic fasting intervals—produce measurable shifts in metabolic markers linked to aging-related processes.

The result is not a fixed protocol but a flexible framework. When periodic metabolic challenge is combined with nutrient-dense eating, regular physical activity, adequate sleep, and stress regulation, physiological systems function with lower metabolic burden. Small adjustments, repeated consistently over time, correspond with differences in metabolic consistency and cognitive performance, supporting a more stable internal environment as aging progresses.

CHAPTER 8

Sleep and Aging: More Than Rest

"Sleep is the golden chain that ties health and our bodies together."
—Thomas Dekker

Amid the relentless momentum of daily life—where productivity often overshadows rest—sleep quietly slips into the background. Society celebrates ceaseless activity, constant connectivity, and tireless ambition, yet emerging research continues to illuminate the essential role of sleep. Far from a passive retreat, sleep is an active biological process that supports physical restoration, mental clarity, and the renewal of key physiological functions.

Historically, sleep was viewed as the body's way of recovering from fatigue. Researchers now describe sleep as a foundational component of long-term well-being, closely tied to cellular maintenance, cognitive performance, metabolic balance, immune activity, and hormonal rhythms. These connections clarify how sleep contributes to human functioning and enhance our ability to adapt and meet the demands of daily life.

Despite its importance, sleep remains one of the most compromised elements of modern living. Stressful commutes, extended work schedules, and evening screen time all contribute to sleep routines that are shorter, lighter, and less restorative than the body may require. Over time, insufficient or irregular sleep patterns can alter how we feel, think, and operate, with consequences that are subtle at first but become more apparent as they accumulate.

In the pages ahead, we will examine how sleep patterns relate to long-term well-being and how restoring healthy sleep can support physical, cognitive, and emotional functioning. By exploring how restorative sleep contributes to physiological balance and overall vitality, we begin to appreciate sleep not only as a necessity but as an intentional act of self-care.

To grasp sleep's full impact, we must examine the processes that take shape each night—processes that influence how we experience the day that follows. What occurs during sleep does not remain isolated there; it carries forward into how we think, feel, and function. To understand sleep's understated yet transformative influence, we must delve into the quiet mysteries that unfold each night, enriching every aspect of waking life.

Understanding Sleep: Foundations of Restoration and Brain Health

Sleep is a familiar albeit complex biological process, essential for supporting physical restoration and cognitive function. Each night, the body and brain move through distinct stages, each contributing to different aspects of recovery. Among these, deep sleep (slow-wave sleep) and rapid eye movement (REM) sleep stand out for their roles in physical renewal, memory processing, and emotional balance. Together, they enable restoration of the body and steadiness of the mind. The structure of these stages helps clarify why sleep exerts such broad physiological and psychological influence.

Deep sleep occurs primarily during the first half of the night and is characterized by slow, high-amplitude brain waves. During this stage, muscles relax, heart rate and blood pressure decrease, and the body undergoes shifts in hormone release. These physiological changes are associated with tissue maintenance, immune activity, and energy regulation. Deep sleep is also central to the function of the glymphatic system—a network that helps clear metabolic byproducts from the brain, including proteins such as beta-amyloid and tau. Research suggests that this nightly clearance contributes to long-term brain health by promoting neural stability and homeostasis. This slow-wave stage plays a central role in sustaining the neurological foundations of daily function.

Sleep disturbances that reduce time spent in deep sleep may influence how efficiently these restorative functions occur, and some research links inadequate time in this stage with markers associated with cognitive decline. Deep sleep also facilitates neuroplasticity—the brain's capacity to reorganize and adapt. During this stage, neurotrophic factors such as brain-derived neurotrophic factor (BDNF) are active, enabling neuronal communication and learning. Memory consolidation, mental clarity, and creative problem-solving are all closely tied to the activity that unfolds during deep sleep. When this stage is shortened or fragmented, the effects often emerge not as dramatic failures but as subtle shifts in clarity, recall, or mental endurance.

In contrast, REM sleep dominates the latter half of the night and is closely linked with the brain's emotional and cognitive work. During REM, neural circuits reorganize, integrating new information and reinforcing important memories while filtering out unnecessary ones. This stage contributes to emotional processing, learning, and mental adaptability—qualities essential for psychological well-being. Its influence often emerges the next day as greater emotional steadiness and improved flexibility in thinking.

The interplay between slow-wave and REM sleep unfolds within a broader temporal system shaped by circadian rhythms. The circadian rhythm, influenced primarily by natural light, coordinates sleep–wake timing, body temperature, hormone levels, and metabolic processes. Disruptions to this system—whether through shift work, frequent travel, or irregular schedules—are associated with changes in mood, metabolic markers, inflammatory responses, and overall physiological balance. When the circadian rhythm falls out of sync, the body often signals its strain through shifts in alertness, mood, or general physiological harmony.

A second regulatory system, sleep homeostasis, governs the build-up of sleep pressure across waking hours. This pressure is shaped in part by adenosine, a neuromodulator that accumulates during wakefulness and promotes sleepiness. When sleep is insufficient or irregular, this balance may be disrupted, contributing to fatigue, reduced cognitive performance, and challenges with emotional regulation. These effects may unfold gradually, becoming noticeable only when the imbalance persists.

By understanding these foundational mechanisms—slow-wave sleep, REM sleep, circadian rhythms, and sleep homeostasis—we gain insight into how sleep influences physical function, cognitive performance, and emotional steadiness. Prioritizing restorative sleep **promotes** essential processes such as glymphatic clearance and neuroplasticity, contributing to long-term adaptability and overall well-being. Sleep, in this sense, becomes not only a biological necessity but a stabilizing force woven through every dimension of daily life.

Hormonal and Circadian Regulation by Sleep

Sleep plays a central role in coordinating hormonal activity that supports internal regulation. When sleep is shortened or disrupted, these regulatory processes can become altered. Inadequate or irregular sleep corresponds with changes in cortisol levels, insulin signaling, and appetite-related hormones such as leptin and ghrelin. Leptin, which signals fullness, and ghrelin, which signals hunger, can fluctuate in ways that influence eating behavior and energy balance. These hormonal changes help explain why sleep quality is closely linked to weight regulation and metabolic balance. Even modest disturbances can increase demands on the body, often appearing as shifts in appetite, energy, or day-to-day stability rather than abrupt dysfunction.

Beyond the hormonal dimension, sleep is closely integrated with circadian rhythms—the body's internal timekeeping system. These rhythms help coordinate sleep—wake cycles, temperature regulation, hormone release, and metabolic timing—signals that support daily stability. Irregular schedules, late-night light exposure, or frequent travel across time zones can disrupt this rhythm, and such disruptions have been associated with changes in mood, metabolic markers, inflammatory responses, and cognitive changes. When circadian cues drift from their natural timing, the effects often surface as shifts in alertness, emotional steadiness, or metabolic balance. Aligning daily routines with one's natural chronotype—whether early bird or night owl—is increasingly recognized as a way to support overall physiological balance.

Sleep also interacts closely with immune function. Restorative sleep supports efficient immune signaling and the formation of immune memory. Research links sleep quality with vaccine responses and with how the body manages infections. Sleep also contributes to outward signs of repair: nighttime growth-hormone activity aids tissue maintenance, including skin repair and collagen production. Sleep and the microbiome influence one another as well—changes in sleep can alter microbial balance, while gut health can affect sleep through the gut—brain axis. These visible changes often reflect the body's internal work of repair and renewal, showing how sleep extends from inner physiology to outward well-being.

Rapid advances in genetics and wearable sleep-tracking technologies have opened the door to personalization of sleep. AI-powered sleep apps and modern wearables offer detailed insights into sleep stages, duration, timing, and variability. Genetic analyses may one day help clarify individual sleep tendencies and preferences. Recognizing these makes it easier to see how sleep influences everyday experience. Making sleep a consistent priority, supported by informed lifestyle habits, becomes an investment in long-term well-being—enriching both nightly restoration and daily experience.

Sleep and the Biology of Aging

Sleep influences many biological processes associated with aging. Key mechanisms—telomere dynamics, DNA maintenance, epigenetic regulation, and mitochondrial function—are all areas of active scientific investigation, and sleep plays a recognized role across these domains. These pathways reveal how nightly rest intersects with the body's long-term processes of repair and adaptation.

Telomeres, protective caps at the ends of chromosomes, naturally shorten as cells divide. Evidence indicates correlations between sleep patterns and telomere length, suggesting possible links between restorative sleep and healthier cellular aging markers. Conversely, chronic sleep disruption has been associated with shorter telomeres in some studies, potentially reflecting greater biological strain. These observations highlight how sleep may intersect with markers of cellular aging. For many researchers, this relationship offers a window into how lifestyle factors and molecular aging may converge.

Sleep also corresponds with DNA-repair activity. During restorative sleep, the body engages in processes that help maintain genomic stability, including pathways such as base excision repair (BER) and nucleotide excision repair (NER). These mechanisms address DNA damage that arises from metabolic activity, environmental exposures, and oxidative stress. Certain repair enzymes, including PARP-1, may be more active during sleep, underscoring sleep's relationship with cellular maintenance. This corrective work does not prevent damage outright, but it helps ensure that accumulated stressors are met with ongoing correction.

Epigenetic regulation—chemical modifications that influence gene expression without altering the underlying DNA—appears sensitive to sleep quality. Insufficient or fragmented sleep has been linked with epigenetic shifts involving metabolism, inflammation, and stress responses. Conversely, consistent, restorative sleep is associated with regulatory activity that aligns with healthier cellular function over time. Sleep functions less as a direct driver and more as a condition that supports the adaptive responses through which genes are expressed.

Mitochondrial health is another focal point. Mitochondria, the cell's energy factories, can undergo stress and decline with age. Sleep appears to influence mitochondrial biogenesis and repair pathways, including signaling proteins such as PGC-1α, which help regulate energy metabolism and mitochondrial adaptation. Although this area continues to develop, available evidence points to a potential relationship between sleep and cellular energy regulation. Rather than indicating a single mechanism, the work to date supports a broader principle: sleep participates in the ongoing adjustments that help sustain biological function.

Sleep and Metabolic Health

Sleep interacts closely with metabolic and cardiovascular processes, coordinating hormonal activity and energy regulation. Irregular or insufficient sleep corresponds with changes in appetite-regulating hormones, insulin signaling, cortisol rhythms, and inflammatory markers. These relationships clarify why sleep duration and quality are closely tied to metabolic regulation and cardiovascular function. Consistent nightly rest supports coordination across multiple physiological systems.

Dietary choices interact with sleep quality and metabolic regulation. Diets rich in lean proteins and dietary fiber correspond with steadier nighttime blood-glucose dynamics and a more balanced gut microbiome, factors that relate to sleep quality. Aligning meal timing with circadian cues—such as eating earlier in the day or keeping evening meals lighter—has been linked to differences in glucose regulation, lipid measures, and inflammatory markers. These observations underscore the close integration of nutrition and sleep in regulating daily energy use. When eating patterns and sleep timing remain consistent, metabolic and cardiovascular systems operate with greater balance.

Sleep and the Microbiome

The human microbiome is a complex community of microorganisms, with its greatest density in the gastrointestinal tract. This microbial ecosystem interacts with multiple aspects of human biology, including mood, metabolism, immune activity, and sleep—wake regulation. Through the gut—brain axis—an active communication network linking the digestive tract and the central nervous system—gut microbes influence emotional and cognitive processes as well as sleep regulation. These interactions illustrate how gastrointestinal physiology integrates with broader regulatory systems in the body.

Certain beneficial bacterial groups, such as *Lactobacillus* and *Bifidobacterium*, participate in metabolic pathways involving dietary tryptophan, which is used to synthesize serotonin, a neurotransmitter associated with sleep regulation and emotional steadiness. The microbiome also contributes to the production of compounds such as GABA, a neurotransmitter involved in calming neural activity. These interactions illustrate how gut health and sleep quality may influence one another. In many cases, the effects emerge gradually, affecting how readily the nervous system settles at night and how it responds to daily demands.

When the microbiome becomes imbalanced—a state commonly referred to as dysbiosis—shifts in inflammatory markers such as IL-6 and TNF-α have been observed, along with changes in stress responses and sleep quality. Dysbiosis has also been linked with various metabolic and neurological conditions, reflecting the interconnected nature of gut health and overall physiology. Poor sleep can further influence microbial composition, creating a cycle in which each domain affects the other. This bidirectional dynamic illustrates how disturbances in one system may reverberate across multiple layers of physiology.

Recognizing the reciprocal relationship between sleep and the microbiome supports a system-level understanding of regulation, in which rest, diet, movement, and emotional regulation interact to influence neuroendocrine, metabolic, immune, and autonomic systems. From this perspective, the microbiome is not a separate domain but an active contributor to how these systems function.

Enhancing Sleep Through Lifestyle: Nutrition, Movement, and Practical Strategies

Lifestyle choices related to nutrition and physical activity influence sleep quality, with effects that become more pronounced with age. Adjustments in these areas can support deeper and more consistent sleep. At a physiological level, these choices affect the conditions that allow the body to enter and sustain nighttime rest.

Nutrition plays an influential role in sleep quality. Lean proteins such as poultry, fish, eggs, and legumes contain amino acids like tryptophan and glycine, both of which are studied for their relationship to relaxation and sleep regulation. High-fiber foods—oats, fruits, vegetables, nuts, and whole grains—are often associated with steadier overnight glucose control, which may help limit sleep disruptions. Conversely, heavy, high-fat, or ultra-processed foods eaten late in the day have been linked to less consistent sleep quality. Healthy fats found in foods such as olive oil, avocados, and nuts—when eaten earlier—may contribute to more stable energy signals that support smoother movement into sleep. These nutritional influences highlight how daytime choices often echo into the night.

Physical activity relates closely to sleep regulation. Resistance training influences growth-hormone dynamics, while aerobic activity interacts with circadian timing. Different forms of movement engage distinct physiological pathways that support the body's shift toward nighttime rest and overnight recovery.

Creating the right environment further enhances this foundation. A cool bedroom temperature is often recommended in sleep research due to its association with the body's natural drop in core temperature and sleep initiation. Reducing evening exposure to blue light from devices helps maintain natural circadian cues. Relaxation techniques—gentle stretching, mindfulness meditation, deep breathing—help ease the movement from wakefulness to rest by calming the nervous system and supporting a consistent nighttime rhythm. These environmental choices form a backdrop against which the body can move more naturally toward sleep.

By combining nutrition, regular movement, and intentional bedtime practices, daily routines create conditions more favorable for restorative sleep. These habits relate to steadier mood, improved daily balance, and a smoother progression into nighttime rest. In this way, ordinary choices made during the day influence the quality of sleep that follows.

The Hidden Architect of Life: Sleep's Role in Our Future

Sleep is far more than a nightly pause—it can be understood as a contributor to long-term clarity, stability, and resilience. Each night, restorative sleep renews both body and mind, replenishing energy and strengthening the capacity to meet the day with focus and composure. It touches countless aspects of biology, relating to cellular maintenance, immune activity, and cognitive performance. Sleep relates closely to emotional steadiness, shaping how daily challenges are met with perspective. In this way, sleep lends the body a form of internal coherence that supports how steadily we move through waking life.

By incorporating practical strategies and personalized approaches—as explored earlier in this chapter—people can engage with sleep in more informed ways. Advances in wearable technology and sleep-tracking tools now offer individualized data that can inform personal habits, helping people identify the routines, schedules, and environments that best align with their own sleep patterns. These tools do not replace lived experience, but they can clarify how small choices across the day influence the ease with which sleep arrives at night.

Consistent sleep is linked to clearer cognition, steadier emotional regulation, and effective physical restoration. Rather than a passive state, sleep functions as an active biological process essential to repair and regulation. Processes engaged during sleep influence how reliably the body maintains function over time, affecting metabolic control, immune activity, and cognitive performance. In this sense, sleep is not separate from daily life but a prerequisite for sustaining it.

CHAPTER 9

Stress, Resilience, and the Aging Mind

"The greatest weapon against stress is our ability to choose one thought over another."
—WILLIAM JAMES

In the quest for longevity, the mind often remains in the shadows, receiving less attention than it warrants. Psychological attributes such as mental steadiness, emotional steadiness, and thoughtful stress management form indispensable pillars of enduring health—quiet in their influence, yet consistently powerful. These internal capacities affect our lives profoundly, influencing how we experience health, navigate challenges, and sustain a sense of meaning.

Mental steadiness—the ability to adapt fluidly to adversity and regain equilibrium—functions like a flexible shield, bending without breaking under life's relentless pressures. Rather than shattering during turbulent moments, the mind navigates difficulty with poise, preserving emotional balance and reinforcing overall stability. In harmony with this steadiness, emotional consistency nurtures optimism, fosters gratitude, and strengthens meaningful social bonds—qualities often associated with greater ease, deeper relationships, and a more grounded experience of aging.

Encouragingly, psychological strength is not limited to those naturally predisposed to it; rather, it can grow within anyone through consistent, accessible practices. Simple, evidence-informed habits such as relaxation exercises, gratitude journaling, cultivating close relationships, and cognitive reframing help build emotional fortitude while easing the burden of daily stress. Integrating these practices into everyday life enriches our experience of the present moment and supports a more grounded, intentional relationship with the years ahead.

The Scientific Foundations of Mental Resilience, Stress Management, and Longevity

At biological and neurological levels, persistent stress exerts wide-ranging effects on the body. While short bursts of stress can sharpen focus and enhance temporary performance, prolonged stress is associated with changes that influence both physical and cognitive well-being. Much of this influence relates to chronically elevated cortisol, the body's primary stress hormone. Sustained cortisol activity has been linked in research to accelerated telomere shortening—the gradual erosion of protective chromosomal caps. As telomeres shorten, cells move toward senescence, a state in which they release signaling molecules associated with inflammation. This cumulative burden is described as allostatic load, the physiological strain resulting from long-term stress exposure.

Chronic stress has also been associated with shifts in mitochondrial function, including reduced efficiency in cellular energy production. These mitochondrial changes can contribute to oxidative stress and disruptions in metabolic equilibrium. Persistent elevations in cortisol can further influence the hypothalamic–pituitary–adrenal (HPA) axis, the body's central stress-response system, in ways that correspond with immune, metabolic, and inflammatory changes.

Neurologically, prolonged stress exposure has been linked with structural and functional changes in key brain regions. Studies have documented associations between chronic stress and reduced hippocampal volume—a region involved in memory and emotional regulation—as well as decreases in brain-derived neurotrophic factor (BDNF), a protein essential for neuronal adaptability, synaptic communication, and learning. These findings help explain why long-term stress is frequently associated with reductions in cognitive performance and emotional dysregulation.

Structured psychological practices interact with neuroplasticity—the brain's capacity to reorganize and strengthen neural pathways. Functional MRI studies indicate that approaches such as mindfulness and cognitive-behavioral methods are linked to reduced reactivity in the amygdala, the brain's primary threat-detection region, and to stronger connectivity within the prefrontal cortex, which supports emotional regulation and decision-making. These observations suggest that certain psychological practices can foster adaptive neural responses under conditions of ongoing stress.

Resilience as a Protective Factor

Resilience acts as a buffer—not by eliminating life's stressors, but by helping people navigate them more effectively. It resembles a muscle that strengthens through facing challenges and integrating the experiences they bring. Those with greater resilience often appear better able to regain emotional equilibrium after disruptions and maintain a steadier stress response. These adaptive capacities are frequently associated with improved emotional functioning and a more stable overall outlook on life.

From a neurological perspective, this capacity has been linked with favorable levels of proteins such as BDNF, which support neuroplasticity, learning, and flexible thinking. These associations point to connections between resilience and the maintenance of cognitive function. Resilience has also been connected with differences in inflammatory responses and with indicators of neural integrity, suggesting a pattern in which psychological steadiness and biological stability often appear together.

The Role of Mind-Body Practices

Mind—body practices—such as meditation, present-moment awareness, diaphragmatic breathing, and gratitude exercises—engage neuroplastic processes through repeated, intentional shifts in attention. By altering how thoughts and emotional responses are processed, these practices reduce stress reactivity and reinforce neural pathways involved in emotional regulation and attentional control.

These forms of attentional training correspond with measurable physiological changes, including increases in heart-rate variability—a marker related to autonomic regulation—and alterations in neural networks involved in affective processing. Emerging work also examines possible links with mitochondrial function, pointing to interactions between psychological engagement and cellular energy processes.

Integrating these methods into daily life can influence habitual patterns of thought, reduce psychological load, and support more adaptive stress responses. While each can be effective on its own, combining them with structured, evidence-informed methods such as cognitive-behavioral techniques often provides additional benefit. They provide practical means to modify thought patterns, manage emotional demands, and respond more effectively to stress. Together, mindfulness practices and cognitive techniques contribute to steadier mental and emotional function, with downstream effects on physical stress systems.

Physical Activity, Relaxation, and Breathing Techniques

Physical activity serves as a means of easing stress, linking bodily movement with emotional steadiness. Even simple activities—morning walks, leisurely cycling, or gentle yoga—have been associated with shifts in mood and overall well-being. Regular movement relates to endorphin activity, contributes to cardiovascular fitness, and is studied for its influence on immune markers and neural adaptability. Consistent engagement in these activities builds stamina and supports greater psychological flexibility.

Relaxation techniques complement the effects of physical activity. Practices such as deep breathing, progressive muscle relaxation, and guided visualization reduce autonomic arousal and support nervous system recovery, providing accessible ways to reduce stress during periods of sustained demand.

Equally important is the establishment of clear boundaries. Boundaries protect mental and emotional resources by limiting unnecessary strain and supporting deliberate, rather than reactive, responses. By preserving time, attention, and emotional energy, boundaries contribute to clearer judgment, emotional capacity, and adaptive functioning.

Cumulative Benefits for Well-Being

When integrated, practices such as mindfulness, gratitude, regular physical activity, boundary-setting, relaxation techniques, and professional support create a constellation of habits that reinforce one another. These practices have been linked with indicators of healthier biological aging, including telomere dynamics, mitochondrial function, and inflammatory signaling. They also appear to foster emotional steadiness and mental agility, supporting clearer thinking and more flexible responses to stress.

Together, these habits form a cohesive framework for long-term well-being. Beyond easing immediate strain, they nurture adaptability and provide a stable foundation for psychological and physical resilience.

The Mind as a Guardian of Health and Well-Being

Mental steadiness and thoughtful stress management impact lifelong wellness. The interplay between mind and body shows that psychological balance corresponds with broader measures of physical health and overall quality of life. Cultivating adaptability and responding intentionally to stressors helps maintain a more balanced internal environment and supports steadier long-term functioning.

Unmanaged stress has been linked to oxidative and inflammatory stress, along with increased vulnerability in cognitive function. In contrast, a steady mindset emphasizes adaptive responses rather than persistent reactivity. Mental flexibility, supportive relationships, regular movement, and mindful awareness interact to create conditions that help preserve functional capacity as individuals navigate sustained demands and change.

Prioritizing emotional balance and mental strength adds depth to life beyond the goal of longevity. Purpose, fulfillment, and supportive relationships often arise from sustained attention to these capacities. Longevity is not an accumulation of years, but a commitment to continued engagement, enduring connection, and emotional steadiness that carries into everyday life.

CHAPTER 10

Emotional and Spiritual Dimensions of Aging

*"It is only with the heart that one can see rightly;
what is essential is invisible to the eye."*
—ANTOINE DE SAINT-EXUPÉRY, THE LITTLE PRINCE

A fulfilling life—rich in depth as well as duration—requires much more than attention to physical health. While discussions of longevity often highlight nutrition, movement, and sleep, deeper psychological and spiritual dimensions shape meaning, stability, and connection. These foundations influence how life is lived, how challenges are met, and how a sense of grounding and purpose takes form.

Emotional steadiness, purposeful living, and meaningful relationships each contribute to a more enriched approach to aging. This chapter examines scientific perspectives on emotional and spiritual wellness, including ways in which inner life interacts with physical processes. It then turns to practical strategies for integrating these dimensions into daily routines, creating a coherent framework for long-term well-being.

The Science of Emotional and Spiritual Health

Emotional and spiritual wellness are not abstract ideas—they interact with biological systems in ways that continue to be examined.

Psychoneuroimmunology, which studies the connections among emotional states, nervous-system activity, and immune function, clarifies how inner experience and physical physiology can intersect.

Positive emotional states—such as gratitude, compassion, and optimism—have been associated with favorable aspects of metabolic regulation, including steadier blood-glucose responses. Positive affect has also corresponded with lower cortisol levels, more balanced immune signaling, and healthier cardiovascular indicators. Higher heart-rate variability (HRV)—a measure of autonomic balance and adaptability—has been observed in people who consistently cultivate forms of positive emotional engagement.

Neuropeptides likewise play an important role in emotional and relational life. Oxytocin, released during meaningful connection and social bonding, has been associated with regulated stress responses and more adaptive activity within the hypothalamic—pituitary—adrenal axis, supporting emotional steadiness during challenging periods.

Neuroscientific observations deepen this picture. EEG and fMRI investigations indicate that mindfulness meditation and contemplative prayer can shift brain-wave activity toward states of calm attention while engaging regions such as the anterior cingulate cortex and prefrontal cortex—areas involved in emotional regulation and executive function. Long-term meditative practice has also been observed alongside greater cortical thickness in regions associated with emotional processing and cognitive control, suggesting potential long-term neural adaptation.

Longitudinal observational work has explored connections between spiritual engagement and various indicators of well-being in older adults. Associations have been reported between regular spiritual or religious participation and lower mortality rates, though causation cannot be inferred. Spiritual practices have also been linked with differences in inflammatory biomarkers such as C-reactive protein (CRP) and interleukin-6 (IL-6). Mindfulness-based programs have been examined for their associations with telomerase activity—a key enzyme involved in telomere maintenance—an area of interest in discussions of cellular aging.

Epigenetic observations add another dimension. Mindfulness and related practices have been linked with gene-expression patterns involving NF-κB, an inflammatory pathway relevant to aging biology.

Creative pursuits provide an additional avenue into emotional and spiritual well-being. Activities such as painting, writing, and music engage dopaminergic reward circuits and support emotional expression. Neuroimaging work indicates that creative involvement can strengthen connectivity within the default mode network, a system important for self-reflection, emotional processing, and internal awareness. Such associations are often described alongside greater emotional stability and cognitive flexibility.

Together, these lines of inquiry suggest that emotional and spiritual life cannot be separated from the biological processes involved in aging. While many questions remain open, the evidence increasingly points to interactions that extend beyond subjective experience into physiological domains relevant to resilience and long-term function.

Emotional and Spiritual Wellness in Longevity Practices

Emotional and spiritual wellness contribute to a purposeful, enduring life. Integrating these dimensions into daily routines involves choosing habits that support psychological adaptability, deepen spiritual connection, and clarify a sense of meaning. Mindfulness and meditation offer accessible pathways toward emotional steadiness and mental clarity. Even brief moments of reflection can ease tension and create a sense of focus. Embedding mindfulness in ordinary experiences—such as conscious breathing, attentive listening, or appreciating nature—turns routine moments into opportunities for restoration.

Practicing gratitude likewise supports emotional balance. Keeping a gratitude journal, acknowledging meaningful experiences, or expressing appreciation through acts of kindness can gently shift perspective and nurture a more open outlook. Strong social connections—formed through shared meals, group movement, meditation gatherings, community involvement, or volunteering—also create pathways for engagement, countering isolation and strengthening emotional cohesion.

For many, consistent engagement in worship contributes meaningfully to spiritual well-being. Sermons, prayer, contemplative rituals, and spiritually resonant music can offer comfort, clarity, and renewal. For those who embrace faith, regular worship may foster a deepening relationship with the divine, providing guidance, emotional grounding, and a strengthened sense of purpose. Such a relationship adds dimension to daily experience, supporting spiritual growth and inner balance.

Integrating these elements into daily living brings emotional and spiritual wellness into the wider work of longevity, shifting attention from the length of life to the experience of living it with depth and presence.

CHAPTER 11

The Pleasure Principle: Sexual Intimacy and Aging

I love you without knowing how, or when, or from where. I love you straightforwardly, without complexities or pride."
—PABLO NERUDA

Sex remains widely misunderstood, often dismissed as mere indulgence or trivialized as fleeting gratification. This narrowed view overlooks its role as a meaningful expression of emotional closeness and physical connection—and the many ways these experiences contribute to overall well-being. Beyond momentary satisfaction, sexual intimacy can steady us emotionally, ease loneliness, and strengthen bonds during life's inevitable challenges. Over time, these influences may accumulate, supporting a shared sense of connection and enriching the experience of long-term partnership.

Sex is more than pleasure—it carries personal and relational significance. Consider intimacy as a natural human experience that nurtures emotional bonds and strengthens connection in subtle yet powerful ways.

In this chapter, we move beyond familiar assumptions, exploring how passion, relational harmony, and physical closeness intertwine. Pleasure becomes a natural and meaningful pathway for supporting emotional vitality and relational depth. By intentionally cultivating sexual intimacy, couples invest in their shared well-being and the richness of their lifelong partnership.

Fire and Foundation: Understanding Sex and Intimacy

Sex and intimacy, though closely linked, serve complementary roles.

Sex is dynamic. It expresses physical desire, merges with emotional affection, and brings immediacy to relational connection. Beyond pleasure, sexual activity is often associated with changes in mood, a sense of relaxation, and greater emotional ease. Many people describe feeling calmer, more present afterward, suggesting that sexual expression contributes meaningfully to both emotional and relational well-being.

Intimacy offers the steadier counterpart, developing gradually through trust, vulnerability, and genuine emotional closeness. It emerges in everyday gestures—a comforting embrace, a shared laugh, or a quiet conversation after a difficult day. These expressions anchor relationships and create stability amid uncertainty.

When sex and intimacy harmonize, their combined influence can deepen connection and evoke neurochemical responses associated with bonding. Their synergy supports companionships marked by warmth, resilience, and satisfaction. Recognizing and nurturing this interplay strengthens emotional bonds and sustains lasting passion.

The Science of Sexual Health: The Hidden Link Between Sexual Intimacy and Well-Being

Sexual experience intersects with many aspects of well-being, contributing to physical vitality, emotional steadiness, and a deeper sense of connection. Like movement or nourishing food, a fulfilling sex life is frequently associated with long-term indicators of wellness.

At the cellular level, sexual activity has been examined for its connections to telomere dynamics—the protective caps on chromosomes that naturally shorten with age. Telomere shortening may be influenced by chronic stress and inflammation. Research published in *Psychoneuroendocrinology* has noted correlations between regular sexual activity and longer telomere length in certain groups, pointing to a potential relationship between intimacy and cellular aging markers. These observations invite further exploration of how intimacy may intersect with processes relevant to cellular aging.

Scientific work has also explored ejaculation frequency in relation to prostate outcomes. The CAPLIFE study, for example, reported that men with higher ejaculation frequency showed lower odds of more aggressive forms of prostate cancer. These findings remain observational and do not imply a preventive effect, but they highlight behavioral associations that merit careful and continued investigation.

Sexual activity is also associated with temporary hormonal shifts. Increases in growth hormone, testosterone, and estrogen can occur during arousal and orgasm. These hormones take part in broader physiological processes—including metabolism, tissue maintenance, and libido—which may contribute to the sense of vitality many people describe after intimate experience.

Cognitively, sexual activity corresponds with increased cerebral blood flow in brain regions involved in memory, emotional processing, and attention. Neurotransmitters such as prolactin, dopamine, and serotonin shift after orgasm and may contribute to changes in mood, emotional comfort, and relaxation.

Some research has also examined connections between sexual activity and immune function. Many people describe a sense of eased tension after intimacy, a response that reflects the release of endorphins—natural compounds involved in modulating discomfort.

When emotional closeness accompanies sexual experience, additional bonding-related chemistry comes into play. Oxytocin and vasopressin—hormones associated with trust and connection—tend to rise more noticeably during partnered intimacy, contributing to feelings of attachment and emotional security.

Sexual intimacy contributes not only to immediate pleasure but also to longer-term emotional and relational well-being. Approached with care and openness, it becomes one way couples can nurture a shared life that feels connected and emotionally supportive across the long arc of being together.

Sexual Intimacy and Sleep: A Meaningful Connection?

One area of research explores the relationship between sexual intimacy and sleep. Sexual activity—particularly partnered intimacy—has been associated with reports of improved subjective sleep quality, with many people describing an easier transition into sleep and a deeper sense of rest following intimacy. These effects are often discussed in relation to physical relaxation, emotional attunement, and the hormonal shifts that follow orgasm.

The presence of a trusted partner can further support this response. Emotional connection and physical closeness contribute to a sense of comfort and safety that promotes relaxation before sleep, and partnered intimacy is frequently linked with more favorable sleep reports than solitary activity.

Restorative sleep supports multiple domains of human functioning, including emotional steadiness, cognitive clarity, and metabolic rhythms. Sexual intimacy does not determine these systems, but it can participate in a reciprocal cycle: intimacy supports rest, rest supports well-being, and well-being helps sustain a more satisfying intimate life.

Life Changes: Sexual Intimacy and Aging

> *"Pleasure is the flower that passes; remembrance, the lasting perfume."*
> —Jean de Boufflers

From youth onward, developing a healthy sense of one's sexuality helps create a foundation for lifelong comfort and ease within intimate relationships. Early conversations about sex and intimacy act like seeds planted for future confidence and clarity.

Midlife brings hormonal changes that naturally alter sexual expression. Men may experience shifts in testosterone, and women often encounter changes in estrogen during menopause. These transitions can influence sexual intimacy, but they also invite opportunities for deeper communication, adaptability, and emotional openness.

Later in life, many people continue to derive meaning from sexual intimacy. Affectionate touch, shared sensuality, and emotional presence enrich daily experience, strengthen companionship, and reinforce feelings of mutual support. By acknowledging and adapting to age-related changes, couples affirm that sexuality remains a dynamic and meaningful aspect of life.

Sexual intimacy remains a vital thread within long-term relationships. While physical expression has its own significance, emotional closeness magnifies the bond between partners, elevating the quality of shared life and reinforcing a sense of connection.

Self-Intimacy: Cultivating Inner Connection and Personal Well-Being

"The relationship with oneself sets the tone for every other relationship."
—Jane Travis

Intimacy begins within. Self-connection allows a rare opportunity to deepen emotional awareness, physical comfort, and personal authenticity. When approached with curiosity and care, self-intimacy can support emotional steadiness, ease tension, and strengthen self-understanding.

Physical self-pleasure, though sometimes misunderstood, can serve as a mindful form of self-care. Many people experience shifts in mood, emotional comfort, and relaxation through the natural release of neurotransmitters such as dopamine, serotonin, and oxytocin during physical self-exploration. Some also notice an easier transition into sleep afterward, a response associated with post-orgasmic hormonal changes.

Self-intimacy extends beyond physical expression. Practices such as supportive self-talk, journaling, reflective solitude, and contemplation nurture inner clarity and emotional stability. These forms of non-physical intimacy reinforce authenticity and empathy, enriching relationships with friends, family, and community.

Together, these pathways can help ease loneliness, support emotional steadiness, and contribute to a richer experience of connection. Self-intimacy is not a substitute for partnered closeness but a companion to it—expanding our capacity for meaningful relationships and deeper human insight.

The Case for Intentional Sex: A Meaningful Investment in Connection

> *"And the day came when the risk to remain tight in a bud was more painful than the risk it took to blossom."*
> —Anaïs Nin

Sexual intimacy offers far more than physical pleasure—it plays a meaningful role in emotional steadiness and a deeper sense of connection. Yet in the pressures of modern life, it often slips to the margins. Stress, digital distraction, and fatigue can quietly draw partners away from one another.

Reframing sexual intimacy as a form of self-care invites couples to approach it with intention rather than urgency. When tended with attention and openness, intimacy can enhance emotional closeness, ease tension, and support a sense of unity between partners. Rather than an indulgence, it becomes an investment in mutual well-being.

Busy schedules and shifting desires can easily sideline intimacy, leaving couples feeling disconnected. Observational research notes that couples who remain engaged with their intimate lives often report stronger emotional bonds and a greater sense of relational satisfaction. Intentional intimacy does not require rigid planning; it rests on presence, communication, and a willingness to remain open with one another.

Couples who prioritize closeness frequently describe a deeper sense of stability and connection. Through intentional intimacy, fleeting moments of pleasure can mature into more enduring sources of shared meaning and companionship.

Prioritizing sexual intimacy enriches relationships, fosters fulfillment, and strengthens the bonds that support long-term partnership. By embracing the erotic dimension of human connection with openness and curiosity, couples engage a powerful and enduring resource—one that can deepen their shared life and bring a richer sense of meaning to the way they share the journey of aging together.

Chapter 12
Hormones and Aging: Timing, Balance, and Biological Change

"The afternoon knows what the morning never suspected."
—Robert Frost

Aging is akin to a delicate dance, choreographed by intricate shifts in our hormonal balance. These invisible chemical messengers govern nearly every facet of our lives—from metabolism and cognitive sharpness to immune coordination and emotional regulation. Although hormonal changes accompany aging, their consequences are not predetermined. By understanding these shifts, making conscious lifestyle choices, and considering appropriate medical evaluation, we can influence how aging is experienced.

Insulin, the hormone that directs blood sugar into cells to meet daily energy demands, plays a central role in glucose regulation. In youth, this system typically functions efficiently. With advancing age, however, cells may become less responsive to insulin's signals, leading to impaired glucose handling and metabolic changes associated with fat accumulation and inflammation. Resistance training and eating patterns rich in fiber and lower in refined sugars are associated with improved insulin responsiveness, helping maintain more consistent glucose regulation into later years.

Cortisol, often dubbed the "stress hormone," functions as both protector and potential disruptor. It acts as the body's alarm system, sharpening focus and mobilizing energy during moments of acute stress—such as narrowly avoiding an accident. Chronic stress, however, keeps this alarm persistently ringing, contributing to prolonged elevations in cortisol. Over time, this state influences inflammation, mood regulation, cognitive performance, and visceral fat accumulation.

Thyroid hormones—particularly T3 and T4—serve as the body's metabolic thermostat, governing energy, weight, temperature, and cognitive clarity. Think of them as skilled conductors, setting the pace of internal biological processes. With age, shifts in thyroid function may appear, influencing metabolism and contributing to fatigue or cognitive fog. Foods containing iodine, selenium, and omega-3 fatty acids provide nutritional input for thyroid-related pathways, working alongside the body's broader physiological systems.

Growth hormone (GH), essential for tissue repair, muscle maintenance, and multiple metabolic processes, contributes to the body's regenerative capacity. Natural GH production often declines with age, coinciding with changes in muscle mass, body composition, and recovery. Lifestyle practices such as regular strength training, adequate sleep, and time-restricted eating have been associated with hormonal signaling patterns that favor GH release, offering non-pharmacological ways these processes may be supported.

Testosterone—important for both men and women—plays roles in libido, muscle strength, bone density, and emotional regulation. With advancing age, changes in testosterone levels in some men (sometimes referred to as andropause), as well as hormonal transitions that affect women differently across the lifespan, are often accompanied by shifts in energy, mood, and physical capacity. Strength training, stress management, and restorative sleep have each been associated with hormonal signaling related to testosterone, reflecting how closely endocrine function is tied to broader daily habits. Nutrients such as zinc and vitamin D also contribute to pathways involved in testosterone metabolism. In situations where a clinically confirmed deficiency is present, hormone replacement therapy is typically evaluated within the context of professional medical care.

Estrogen plays a significant role in bone biology, cardiovascular physiology, and cognitive function. In women, the transition into menopause brings a rapid decline in estrogen, a change widely discussed in the literature in relation to shifts in bone density and cardiometabolic indicators. Foods containing phytoestrogens—such as flaxseeds and soy—along with vitamin D and strength training, are often incorporated into lifestyle approaches that help maintain general physical function. FDA-approved bioidentical hormone therapy may be considered for women experiencing particularly challenging symptoms, with decisions guided by individualized evaluation. In men, smaller amounts of estrogen produced through testosterone conversion also contribute to physiological balance, underscoring its relevance across sexes.

DHEA (dehydroepiandrosterone), a precursor to hormones such as testosterone and estrogen, participates in multiple regulatory pathways involving energy, stress response, and immune coordination. Its levels decline gradually with age, much like daylight fading at sunset, and lower concentrations have been associated with changes in energy and emotional balance. Lifestyle practices that promote metabolic and emotional resilience—regular exercise, restorative sleep, and balanced nutrition—help support endocrine environments in which DHEA-related pathways function more effectively.

Melatonin acts as a critical signaling molecule aligning the body's circadian rhythms with environmental cues, particularly darkness. By coordinating these rhythms, melatonin contributes to restorative sleep and the timing of numerous biological processes. Melatonin production decreases with age, often accompanying disruptions in sleep patterns. Strategies such as reducing evening blue-light exposure—much like dimming lights before a theater performance—maintaining consistent sleep schedules, and creating calm nighttime environments can help reinforce natural rhythms. In clinical discussions, melatonin supplementation is sometimes referenced in relation to sleep support.

Each hormone operates within an interconnected system, influencing not only biological aging but how its effects are experienced over time. Hormonal change is an inherent part of the life cycle, though its impact varies with awareness and response. Understanding how these signals interact allows for a range of approaches, including mindful movement, nourishing food choices, meditative practices, and professionally supervised interventions. Together, these approaches help sustain emotional steadiness, cognitive clarity, and continued engagement with life.

Nutrition as the Foundation of Hormonal Balance

Our dietary choices supply the raw materials the body uses for hormone production, influencing everything from metabolism to emotional regulation. Omega-3 fatty acids—found in fatty fish, flaxseeds, and walnuts—have been studied for their role in inflammatory balance and their involvement in pathways related to cortisol and reproductive hormone signaling. Zinc, present in shellfish, pumpkin seeds, and legumes, contributes to biochemical processes involved in immune function and hormonal pathways, including those related to testosterone. Magnesium—abundant in leafy greens, nuts, and whole grains—plays a part in adrenal and neuromuscular activity, affecting the body's response to stress. Selenium, sourced from Brazil nuts, eggs, and seafood, participates in key reactions within thyroid hormone metabolism.

Beyond individual nutrients, more stable blood-sugar dynamics help maintain broader hormonal function. Reducing intake of refined sugars and highly processed foods can improve insulin responsiveness, while fiber-rich diets featuring phytonutrient-dense fruits, vegetables, and whole grains help nourish the gut environment. Because the gut and endocrine systems communicate extensively, a healthy microbiome contributes to many aspects of hormonal signaling.

The Role of Exercise in Hormonal Support

Movement plays a central role in hormonal regulation, shaping how the body manages energy, stress, and physical control. Resistance training—whether using weights or bodyweight exercises—provides mechanical and metabolic signals that contribute to muscle maintenance and metabolic activity. Within exercise physiology and endocrine research, interactions with testosterone and growth hormone pathways have been studied in relation to strength and recovery.

Cardiovascular activities such as running, cycling, or swimming have been associated with improvements in insulin sensitivity and shifts in cortisol dynamics, often through effects on body composition and stress processing. Observational work also notes that very high training loads without sufficient recovery are associated with sustained cortisol elevation, which may blunt some of these effects. Many people report that combining resistance work, aerobic movement, and flexibility practices such as yoga or Pilates provides a sustainable foundation for long-term hormonal and physical adaptation.

The Essential Role of Sleep in Hormonal Regulation

Deep, restorative sleep is fundamental to hormonal function. During restful sleep, the body undertakes a range of regenerative processes—much like a night crew carrying out essential maintenance. These processes influence tissue repair, muscular recovery, metabolic regulation, and the timing of hormonal signals. Melatonin, the hormone associated with sleep onset, helps coordinate circadian rhythms and align numerous clock-regulated systems throughout the body.

Insufficient or inconsistent sleep, by contrast, can affect cortisol cycles, insulin responsiveness, and appetite-related signals such as leptin and ghrelin. These shifts may contribute to changes in energy, cravings, and metabolic control. Establishing consistent bedtime routines, reducing evening blue-light exposure, and creating calming sleep environments can help reinforce stable sleep—wake rhythms and maintain the body's natural hormonal cycles.

Managing Stress to Support Hormonal Stability

Chronic stress acts like an unwelcome guest that overstays its welcome, consistently disrupting physiological balance through prolonged cortisol elevation. While short-term cortisol spikes are essential, sustained elevations influence metabolism, immune function, and emotional regulation. Mindfulness, meditation, yoga, and breathwork offer calming anchors that help the nervous system settle after repeated activation. Fulfilling activities—such as quiet time in nature, creative expression, or gratitude practices—counter persistent stress and reinforce emotional resilience.
Intermittent Fasting as a Metabolic and Hormonal Tool

Intermittent fasting offers a structured way to examine metabolic and hormonal responses, much like periodically rebooting a computer for improved clarity and focus. The widely practiced 16:8 fasting approach—fasting for 16 hours followed by an 8-hour eating window—has been examined for its associations with insulin responsiveness, inflammatory signaling, and hormonal coordination. However, intermittent fasting does not suit everyone. Those experiencing heightened stress, adrenal sensitivity, or increased physiological demand may find that fasting introduces additional metabolic strain. As with tailoring clothing for the right fit, fasting practices benefit from alignment with personal stress levels, daily activity, and overall health so they work with, rather than against, the body.

Targeted Supplementation for Hormonal Support

Thoughtfully chosen supplementation can complement broader approaches to hormonal function—much like adding the right seasoning to enhance a meal. Ashwagandha, an adaptogenic herb, has been studied in relation to stress-related signaling and cortisol regulation. Maca root, long used in traditional settings, has been associated with energy and libido. Omega-3 fatty acids, sourced from fish oil or algae-based supplements, participate in inflammatory and cardiovascular networks that intersect with hormonal activity. Vitamin D contributes to multiple biological processes, including those related to mood, immune activity, and elements of hormone synthesis, and limited sun exposure is common in many populations. Consulting a healthcare professional can help ensure that supplementation choices align with individual needs and safety considerations.

Hormone Replacement Therapy: A Closer Look Ahead

For those experiencing notable hormonal shifts related to aging or specific medical circumstances, hormone replacement therapy (HRT) is one approach discussed within clinical care. The next chapter examines HRT by reviewing relevant research, key considerations, and the contexts in which decisions are made as part of a broader approach to navigating hormonal change.

Emerging Therapies and Future Directions in Hormonal Health

Advances in molecular biology are opening new avenues for investigating how hormonal signaling changes over time. Rather than replacing existing systems, many of these developments focus on interacting with endogenous regulatory pathways, expanding the range of biological questions that can be explored in aging research. From peptide-based compounds designed to engage hormone-related signaling to emerging gene-editing technologies, this area of inquiry continues to evolve. At the same time, wearable technologies are advancing to provide finer-grained insight into metabolic patterns, stress responses, and circadian organization, supporting more detailed observation of physiological variation.

Peptide therapies are among the more actively examined areas of current biomedical research. Peptides—short chains of amino acids—interact with the body's existing hormonal signaling systems, often by influencing endocrine pathways rather than supplying hormones directly. For example, sermorelin acts at receptors in the pituitary gland involved in growth hormone regulation. Other peptides, including CJC-1295 and ipamorelin, are being studied in relation to recovery, energy, and aspects of metabolic regulation. These compounds remain investigational rather than established therapies, but they reflect growing interest in approaches that engage the body's regulatory systems rather than replacing them.

Gene therapy represents another area of active scientific exploration. Tools such as CRISPR-Cas9 allow investigators to study how modifying specific genes might influence hormonal pathways, including those involved in insulin signaling, growth-hormone regulation, and stress responsiveness. Much of this work remains experimental, focused on clarifying foundational mechanisms rather than clinical application. Even so, the possibility of altering genetic influences on hormonal aging—sometimes compared to editing code within biological systems—continues to motivate ongoing research.

Wearable technologies are increasingly shaping how physiological signals related to hormonal regulation are monitored and interpreted in daily life. Continuous glucose monitors provide real-time visibility into glucose fluctuations, offering insight into metabolic responses. Heart-rate-variability trackers reflect aspects of stress-related physiology, while sleep-tracking devices capture patterns related to circadian rhythms. As innovation progresses, future wearables may attempt non-invasive sensing of biomarkers such as cortisol, estrogen, or testosterone through sweat or saliva. Combined with AI-driven analytics, these tools may enhance awareness of physiological trends and connect ongoing monitoring with evolving scientific understanding.

The convergence of these therapies and technologies points toward a more individualized and forward-looking understanding of hormonal health. Rather than focusing only on responses to symptoms, emerging models increasingly emphasize earlier recognition of physiological change and interpretation informed by ongoing measurement and context.

With innovation, however, come ethical and practical considerations. The long-term effects of gene editing and peptide-based interventions require continued study, and questions of equitable access remain central. The possibility of unregulated use underscores the importance of thoughtful oversight, clear ethical standards, and rigorous long-term evaluation.

Advances in gene editing, wearable monitoring, peptide signaling, and related fields continue to expand the scientific study of hormonal regulation and aging. While much of this work remains investigational, it reflects a growing effort to understand aging as a dynamic biological process rather than a fixed trajectory of decline. Over time, these lines of inquiry may help shape how biological change across the lifespan is interpreted, without assuming that every insight must become an intervention.

Hormonal Balance and the Science of Aging Well

Hormones play a central role in aging, influencing energy, metabolism, cognitive function, and emotional regulation. Although hormonal change is often treated as an unavoidable consequence of growing older, it does not require a passive stance. Greater understanding allows people to respond more deliberately—shifting attention away from resisting time and toward sustaining clarity and functional capacity across the years.

Hormonal balance is not limited to symptom management. It influences movement, mental engagement, and the ability to adapt as physiological demands evolve. Nutrition, physical activity, restorative sleep, and stress regulation each contribute to this balance. Ongoing research has expanded this perspective, including clinically guided hormone therapies and other individualized approaches considered in specific contexts.

Hormones are not merely chemical messengers; they shape the background conditions of daily life—how energy is sustained, attention is held, and recovery unfolds. When regulation is disrupted, the effects are often felt across multiple domains. Attending to hormonal health is therefore less an effort to resist aging than an ongoing process of adjustment—one that, for many, supports steadier functioning and sustained engagement as life continues to change.

CHAPTER 13

Hormone Replacement Therapy: Evidence, Risks, and Judgment

"The body is like a piano, and happiness is like music. It is needful to have the instrument in good order."
—HENRY WARD BEECHER

Hormones play a central role in aging, influencing energy, metabolism, cognition, and emotional regulation. Although hormonal change is often treated as an unavoidable consequence of growing older, it does not require a passive stance. With greater understanding, people can respond more deliberately—shifting attention away from resisting time and toward maintaining clarity, function, and responsiveness across the years.

Hormonal balance extends beyond the management of isolated symptoms. It influences how the body moves, how attention is sustained, and how readily adaptation occurs as physiological demands change. Nutrition, physical activity, restorative sleep, and stress regulation each contribute to this balance. Ongoing research has broadened this perspective further, including clinically guided hormone therapies and other individualized approaches that may be considered in specific contexts.

Hormones are not just chemical messengers; they set the background conditions of daily life—how energy rises and fades, how focus is maintained, how recovery unfolds. When regulation is disrupted, the effects are often felt quietly at first, then more clearly over time. Attending to hormonal health is therefore less an effort to defy aging than an ongoing process of adjustment—one that, for many, allows steadier functioning and sustained engagement as life continues to change.

Understanding Hormone Replacement Therapy

HRT can be delivered in several forms, each with distinct pharmacokinetic features. Oral tablets offer convenience but undergo first-pass metabolism in the liver, producing metabolic effects that differ from non-oral routes. Transdermal patches and topical preparations avoid this pathway and can provide more stable systemic exposure. Subcutaneous pellets release hormones over longer intervals, reducing variability between doses. Injections—often used in testosterone therapy—produce cyclical peaks and troughs that may be appropriate in some physiological contexts.

With appropriate medical oversight, HRT can be incorporated into aging-related care hormonal stability. Ongoing research continues to clarify where HRT is most useful and how risks and benefits differ across settings, situating it within the broader scientific inquiry into longevity.

Hormone Replacement Therapy for Women: Tailoring Treatment for Optimal Health

For women in menopause and in the years surrounding it, hormone replacement therapy (HRT) can help maintain more consistent hormone levels. Most regimens use estrogen and progesterone, with testosterone used selectively, even as aging affects many hormones simultaneously. Changes in one hormone can alter how others function, creating interconnected shifts that influence symptoms and daily experience. Recognizing these interactions allows therapy to be matched to specific clinical concerns rather than approached as a single-hormone solution.

Estrogen, Progesterone, and Testosterone

Estrogen Replacement Therapy (ERT) is often discussed in clinical settings in relation to menopausal symptoms such as hot flashes, night sweats, vaginal dryness, and changes in mood. Estrogen's influence extends beyond these symptoms; it contributes to bone strength, vascular responsiveness, and other physiological processes that shift during the menopausal transition. Research has explored how the timing of therapy initiation relates to comfort, daily functioning, and bone maintenance during the postmenopausal period.

After menopause, the balance among estrogen forms changes as estrone (E1) becomes more prominent. Higher estrone levels have been associated with certain estrogen-responsive conditions, and many clinicians use bioidentical estradiol (E2), the form produced before menopause. For women with a uterus, estrogen is typically paired with progesterone to counteract its effects on the uterine lining. Progesterone has also been associated with more stable sleep patterns and emotional regulation in some women.

Testosterone plays a meaningful role in women's physiology, contributing to libido, mood, muscle tone, and cognitive motivation. Levels decline gradually after age 30 and fall further during menopause, which for some women corresponds with fatigue, reduced sexual interest, or decreased physical strength. Testosterone can be delivered through gels, creams, or subcutaneous pellets. In clinical use, testosterone therapy can affect energy, mood, and musculoskeletal performance, though responses vary. Excessive levels can lead to acne, unwanted hair growth, or vocal changes, underscoring the importance of careful dosing and monitoring in clinical contexts.

DHEA (dehydroepiandrosterone), a precursor for both estrogen and testosterone, participates in metabolic activity, stress response, and immune function. As levels decline with age, some women describe lower energy, shifts in emotional regulation, or reduced adaptability during periods of increased demand. Low-dose DHEA supplementation is discussed in some clinical contexts, though responses are variable and higher doses can produce androgenic effects.

Intrarosa and Local Options

Intrarosa is a locally administered therapy discussed in clinical practice for postmenopausal vaginal changes. It contains prasterone (DHEA), which is converted within vaginal tissues into active hormones, distinguishing it from systemic hormone therapies. As a non-estrogen local approach, it is often considered within clinical care as an alternative to systemic estrogen exposure for vulvovaginal symptoms following menopause.

A Holistic Approach to Women's Hormonal Health

Hormone replacement therapy (HRT) can play a role for women navigating the shifts that accompany hormonal decline, but lasting steadiness rarely comes from a single intervention. Hormones behave more like signals moving through a dense network than isolated levers, and their effects depend on the condition of the tissues that receive them. Daily habits—how muscles are used, how stress is absorbed, how sleep unfolds, and how food is metabolized—each influence the fidelity of those signals.

Strength training, for example, alters muscle signaling and glucose handling, giving the body a more responsive metabolic base. Many women notice a clearer sense of physical capacity when these pathways are engaged. Nutrient-dense eating plays a similar role. Proteins, varied vegetables, and healthy fats help stabilize the metabolic context in which hormones exert their influence, making energy levels less erratic across the day.

Stress practices operate on a different but related axis. Cortisol, the hormone most closely tied to pressure and vigilance, sets the background tone for many other hormonal signals. When cortisol rises, the internal system tightens; when it recedes, other signals gain room to register. Practices such as steady breathing, brief pauses during demanding days, or structured meditation can help modulate this background, reducing the interference that often accompanies chronic stress.

Assessment and re-evaluation provide another point of orientation. Hormonal data do not dictate a single course of action, but they reveal how therapy is unfolding and whether adjustments may be warranted. These evaluations function less as instructions and more as reference points, allowing both patient and clinician to understand how various signals are interacting.

For women pursuing hormonal consistency, working with a clinician who understands both physiology and the lived experience of menopause can clarify the process. Decisions about bioidentical hormones, selected supplements, and supportive lifestyle measures emerge from this partnership. Each choice influences how the hormonal network responds, how symptoms evolve, and how the body adjusts to this stage of life.

Hormone Replacement Therapy for Men

As men age, hormonal shifts can influence energy, mood, strength, and cognitive regulation. Testosterone functions less as a singular "driver" and more as a signal moving through multiple systems at once—muscle, bone, metabolism, and the brain's regulatory circuits. When levels begin to fall, often as early as the thirties, the change can register across these systems. Many men notice slower recovery after exertion, reduced sexual interest, or a gradual softening of concentration. These changes unfold at different rates, but they reflect a common underlying decline in androgen signaling.

Hormone replacement therapy (HRT) offers a structured way to respond to these shifts. By restoring testosterone to a physiologic range, therapy can influence muscle protein synthesis, red blood cell production, and aspects of mood and motivation that depend on androgen activity. The goal is not to reverse age, but to stabilize the internal signals that shape how the body responds to daily demands. For some men, this steadier hormonal environment can make physical effort feel more accessible and mental engagement easier to maintain.

Working with a clinician allows these effects to be observed over time. Laboratory assessment, symptom patterns, and changes in function help clarify whether adjustments are needed. This partnership grounds therapy in physiology rather than expectation, giving men a clearer sense of how hormonal change intersects with lived experience.

Testosterone Replacement Therapy (TRT)

Among hormonal interventions for men, testosterone replacement therapy (TRT) has been examined more extensively than any other. Low testosterone alters how signals move through muscle, bone, metabolism, and the brain's regulatory circuits. When those signals weaken, men often experience the change in tangible ways: slower recovery after exertion, reduced sexual interest, shifts in strength, or a narrowing of mental drive. TRT works by restoring androgen signaling to a physiologic range, giving these pathways a clearer channel through which to operate.

As testosterone rises toward a more consistent baseline, several systems respond. Muscle tissue becomes more efficient at repair following strain. Fat metabolism shifts, often changing how the body manages energy. Red blood cell production increases, influencing oxygen delivery and endurance. Testosterone may influence memory, focus, and emotional regulation, reflecting the hormone's reach into neural circuits governing motivation and attention. Early work continues to examine how these pathways interact with aging in the brain, though much remains under investigation.

Testosterone replacement therapy (TRT) can be administered through several clinical delivery methods, including transdermal gels, injections, patches, and subcutaneous pellets. These approaches differ in how testosterone levels fluctuate over time, with some producing steadier exposure and others more variable patterns. In clinical settings, follow-up is used to assess how an individual responds within their broader medical context. Ongoing monitoring—such as tracking blood parameters, fluid balance, and symptom patterns—provides information about how therapy is progressing and whether adjustments may be considered.

Earlier concerns about a strong connection between TRT and prostate cancer have evolved as newer findings have emerged. Current evidence suggests a more complex interaction between testosterone levels and prostate biology, shaped by androgen receptor activity, tissue sensitivity, and age-related changes within the gland. This area continues to develop, and clinicians interpret emerging data with attention to both risk and biological nuance.

For many men, discussions around TRT focus less on reversing age and more on supporting physiological conditions that tend to change over time. Its role in care is often framed in this context. By influencing testosterone availability within established clinical parameters, TRT may affect systems related to muscle maintenance, metabolism, and aspects of cognitive and motivational function as hormonal regulation shifts with age.

DHEA and Growth Hormone: Supporting Energy and Recovery

Testosterone functions within a broader hormonal network, and one of its key intermediaries is DHEA (dehydroepiandrosterone). DHEA serves as a precursor for both testosterone and estrogen and participates in pathways related to metabolism, stress physiology, and recovery. As DHEA levels decline with age, some men report changes in energy, libido, or recovery that coincide with broader shifts in hormonal regulation. When used under clinical supervision, DHEA supplementation may alter downstream hormone levels, though responses vary considerably. Because DHEA feeds into multiple hormonal pathways, careful oversight is required to maintain effects within a physiologic range.

Growth hormone (GH), along with insulin-like growth factor 1 (IGF-1), plays an important role in tissue repair and metabolic regulation. GH is involved in muscle remodeling, fat utilization, and changes in body composition that occur over time. Production declines steadily with age, a shift often accompanied by changes in recovery, strength, and body composition.

Direct GH therapy remains a subject of debate because of potential side effects, and much current research has focused instead on agents that act upstream in the GH axis, such as investigational releasing peptides including sermorelin and CJC-1295. These compounds are studied for their ability to stimulate endogenous GH secretion through pituitary signaling, though they remain outside established clinical practice.

Lifestyle factors also interact with these pathways. Resistance training, higher-intensity exercise, and consistent, restorative sleep are associated with changes in GH signaling and broader endocrine responses. While such practices do not substitute for clinical treatment, they contribute to the physiological context in which aging-related metabolic demands are managed.

Holistic Hormonal Health for Men: Beyond Therapy

Understanding hormonal change begins with recognizing how physiology, medical history, and daily demands interact over time. Testosterone is one component within a broader hormonal network that also includes DHEA, growth hormone, thyroid hormones, and cortisol. These systems do not operate independently; together they influence energy balance, cognitive function, metabolic regulation, and the body's response to physical stress. Considering them collectively offers a more complete view of how strength and functional capacity are maintained as aging progresses.

New tools—including advanced laboratory testing, wearable technologies, and emerging forms of biomarker monitoring—have expanded the ability to observe physiological signals that were previously difficult to track. Used descriptively rather than prescriptively, such data can illustrate how factors like sleep, stress, exertion, or clinical intervention correspond with internal change. When interpreted within clinical care, these measurements help distinguish transient variation from more durable shifts and support more informed discussion.

Medical interventions are typically considered alongside daily habits. Physical activity, nutrient-dense eating, restorative sleep, and stress regulation each influence metabolic and endocrine processes that evolve with age. Together, these factors shape the physiological background against which clinical care is evaluated, placing hormonal signals and physical responses in perspective.

When clinical care, observational data, and stable routines are considered together, age-related change becomes easier to interpret. Hormonal fluctuations can be read with greater clarity, physical demands anticipated more accurately, and shifts in cognitive focus understood as part of an evolving biological landscape. Aging does not eliminate these capacities, but it alters the conditions under which they operate. Recognizing those conditions allows adjustment to remain informed rather than reactive as change unfolds.

Bioidentical vs. Synthetic Hormones

Hormone replacement therapies are often used to promote greater hormonal consistency, but understanding the difference between bioidentical and synthetic hormones is essential. Bioidentical hormones share the exact molecular structure of the hormones the body produces, allowing them to bind receptors and signal in ways that more closely resemble endogenous physiology. Synthetic hormones, by contrast, contain deliberate structural alterations, and those variations can change how they interact with receptors and how they are metabolized by the body.

Conjugated equine estrogens (CEE) illustrate this distinction. Derived from the urine of pregnant mares, they contain several estrogenic compounds that differ structurally from human estradiol, and these differences can influence physiological effects. Certain synthetic progestins follow a similar pattern; because their structure does not mirror natural progesterone, some studies have reported distinct biological responses and side-effect profiles.

The distinction rests on molecular configuration rather than origin. A hormone sourced from plant material may still be synthetic if its structure does not match human hormones. Conversely, a compound created entirely in a laboratory can be bioidentical if its molecular arrangement is identical to what the body produces. Structure—not source—largely determines whether a hormone behaves like its natural counterpart.

FDA-Approved and Compounded Bioidentical Hormones

Bioidentical hormones are available in two main forms: FDA-approved formulations and compounded preparations. FDA-approved options undergo formal review for quality, dosing accuracy, and manufacturing practices. Produced under regulated conditions, these formulations—such as estradiol patches, progesterone capsules, and testosterone gels—must meet predefined benchmarks for purity and consistency.

Compounded bioidentical hormones are prepared by compounding pharmacies for individual patients to accommodate specific dosing needs or delivery formats. This approach allows for options such as creams, lozenges, and injections that are not always available in commercially manufactured products. Because compounded formulations are produced on a patient-specific basis rather than through standardized manufacturing, they are not FDA-approved and do not undergo the same level of large-scale testing or regulatory review. This individualized process can introduce variability in potency or composition, which is why major medical organizations emphasize careful evaluation when comparing compounded preparations with regulated alternatives.

Common Myths About Hormone Replacement Therapy

Few topics in midlife health generate as much uncertainty as hormone replacement therapy (HRT). A persistent concern is the belief that HRT inevitably causes cancer—a view rooted in findings from the 2002 Women's Health Initiative (WHI). That study linked specific hormone formulations—most notably conjugated equine estrogens (CEE) and certain synthetic progestins—to altered profiles of breast, cardiovascular, and cerebrovascular risk. Subsequent analyses have shown that these risks vary substantially according to hormone type, dose, timing, route of administration, and baseline health characteristics.

Bioidentical formulations such as estradiol and micronized progesterone differ structurally from synthetic hormones, and researchers have examined how these differences shape physiological responses. Estrogen therapy alone shows distinct effects in women who have undergone hysterectomy, underscoring the importance of clinical context. In combined therapy, variation among progesterone and progestin formulations can produce different effects on breast and uterine tissues. Rather than reflecting a single risk profile, HRT involves an interaction among genetics, baseline medical factors, lifestyle, and the specific regimen used, underscoring the need for careful interpretation.

A second misconception is that hormone therapy is useful only for severe menopausal or andropausal symptoms. While relief from hot flashes, sleep disruption, and mood changes is often central, hormones also participate in bone turnover, vascular responsiveness, and metabolic regulation. Research on cognition continues to explore possible links between estrogen therapy and aspects of memory and attention, particularly when treatment begins during the early postmenopausal window.

Men experiencing age-related declines in testosterone may notice gradual shifts in libido, strength, energy, and concentration. Testosterone replacement therapy (TRT) has been studied in relation to these changes and is typically evaluated in clinical settings when deficiency is clearly documented, as inappropriate dosing can elevate red blood cell counts, affect prostate physiology, or alter cardiovascular parameters. Careful monitoring is therefore central when TRT is considered in the context of androgen decline.

The practice of hormone therapy continues to evolve alongside advances in precision diagnostics. Biomarkers such as sex hormone—binding globulin (SHBG) and, in some settings, genetic markers like CYP19A1 can provide additional insight into hormone metabolism and response, informing more individualized clinical interpretation.

New delivery methods continue to be examined within clinical research. Transdermal estradiol allows absorption through the skin, sublingual progesterone provides an alternative to oral capsules, and long-acting testosterone pellets offer a sustained release profile sometimes favored in clinical practice. Peptide-based agents such as CJC-1295 and sermorelin are studied for their effects on endogenous growth-hormone signaling, representing a mechanism distinct from direct hormone administration. Collectively, these developments reflect ongoing efforts to better align delivery methods with individual physiology.

As evidence continues to accumulate, clinical approaches are refined and longer-term outcomes become clearer. The broader direction of hormone research emphasizes precision, careful risk assessment, and attention to maintaining functional capacity across later decades of life.

The Path Forward: Maximizing Health Through Hormonal Balance

Aging is universal, but how it unfolds varies with physiology, environment, and personal circumstances. Hormone replacement therapy (HRT) is not a cure for aging, nor does it address the process as a whole. Instead, it is one of several clinical approaches used to address how age-related hormonal changes influence biological regulation over time. These changes typically develop gradually, and understanding their trajectory can inform how aging is interpreted rather than simply endured.

The clinical use of hormone therapy emphasizes specificity. Decisions around HRT are based on medical history, hormonal patterns, and individual tolerability, not uniform expectations. In women, estrogen and progesterone formulations have been studied in relation to menopausal symptoms and physiological changes. In men, testosterone therapy is evaluated only when deficiency is clearly documented and requires careful monitoring because of its systemic effects. Outcomes vary widely, underscoring the importance of cautious interpretation rather than generalized conclusions.

Hormone therapy is typically considered alongside broader aspects of daily life that also influence endocrine regulation. Physical activity, nutrition, sleep, and stress exposure each affect metabolic and hormonal signaling as aging progresses. Together, these factors shape the physiological context in which hormonal measurements and clinical decisions are assessed, rather than acting as adjuncts to therapy.

Ongoing advances in diagnostics and pharmacology continue to refine how hormonal change is studied and monitored. Improved laboratory tools and greater understanding of hormone metabolism allow clinicians to interpret variation more precisely and to weigh potential risks within an individual clinical context. Contemporary research focuses less on altering aging and more on clarifying how biological systems adapt.

Within this framework, hormone therapy represents one way clinicians examine and respond to age-related hormonal variation. Its role is defined by careful monitoring, realistic expectations, and recognition of uncertainty. Rather than promising preservation or enhancement, HRT illustrates how modern medicine approaches aging as a process to be understood, interpreted, and managed.

Part III

Rewriting the Rules of Aging—Longevity Science and Biohacking

"The reasonable man adapts himself to the world; the unreasonable one persists in trying to adapt the world to himself. Therefore all progress depends on the unreasonable man."
—George Bernard Shaw

Biohacking is often associated with bold self-experimentation, but the field of longevity reaches far beyond personal trials. It includes independent exploration alongside structured scientific research, regulatory review, and measured incorporation into medical care. As advanced therapies move from early testing toward broader availability, they bring questions about safety, public health, and ethical responsibility. This section examines those concerns and considers how new interventions may influence the way aging is understood in practice rather than theory.

Biohacking remains a visible part of this movement, offering tools that allow people to take a direct role in maintaining their capabilities. Longevity science, however, extends beyond personal experimentation and focuses on therapies supported by rigorous inquiry and clinical evaluation. Its emphasis lies in findings that are tested, reproducible, and slowly entering medical practice. Distinguishing these approaches is important, because self-directed enhancement and research-based treatment carry different expectations, limits, and levels of oversight. Clarity about this distinction helps keep exploration responsible and grounded in what is known rather than assumed.

Much like pioneers venturing into uncharted territories, biohackers tread carefully between the promise of discovery and the uncertainties of experimentation. Self-testing can offer unique personal insights but also carries inherent risks, particularly when novel treatments lack extensive long-term safety data or clearly defined clinical frameworks. Optimizing health should never come at the expense of well-being. Navigating this complex and swiftly changing terrain demands discernment, solid scientific knowledge, and collaboration with trusted medical and longevity experts. By balancing innovation with prudence, those aiming to extend their healthspan can ensure their journey remains forward-looking, safe, responsible, and anchored in rigorous science.

CHAPTER 14

Rapamycin and Rapalogs: Inside a New Era of Longevity Research

"The only way to discover the limits of the possible is to go beyond them into the impossible."
—ARTHUR C. CLARKE

Across centuries, humankind has pursued ways to delay, halt, or even reverse aging. Myths of mystical fountains and legendary elixirs have long captured the imagination, but rapamycin stands apart—not for folklore, but for the depth of scientific investigation surrounding it. Originally used to prevent organ rejection in transplant recipients, rapamycin has since drawn widespread attention as one of the most closely studied discoveries in modern longevity research. Today, investigators are examining whether it influences not only lifespan, but also healthspan—the years marked by vitality and reduced burden of chronic disease.

At the center of rapamycin's potential is its ability to interact with a fundamental biological system: the mTOR pathway, which governs whether cells prioritize growth or restoration. By selectively inhibiting a key component known as mTOR Complex 1 (mTORC1), rapamycin shifts cellular activity toward repair. This shift activates autophagy—the body's internal recycling mechanism—allowing cells to clear damaged components and maintain internal stability. The resulting cellular state resembles that induced by caloric restriction, an intervention known to affect lifespan and healthspan in multiple species. Because it acts on pathways the body already uses to regulate repair and adaptation, rapamycin offers a pharmacological way to engage cellular programs that are normally activated during periods of nutrient limitation.

Clearing the Wreckage: Rapamycin's Role in Inflammaging, Senescence, and Metabolic Health

Building on its role in cellular repair, rapamycin also interacts with two biological processes closely tied to aging: chronic inflammation and cellular senescence. As the immune system grows older, it often settles into a low-grade, persistent inflammatory state—known as inflammaging—which contributes to conditions such as cardiovascular disease, neurodegeneration, and metabolic decline. Rapamycin influences this tendency by modulating inflammation-related signals and promoting more regulated immune activity in aging tissues.

Senescent cells—damaged cells that stop dividing but continue to release disruptive chemical cues—pose a different kind of challenge. These cells accumulate in tissues over time, affecting nearby healthy cells and accelerating local dysfunction. In research settings, rapamycin has been observed to reduce indicators of senescent-cell burden and the inflammatory compounds they release, providing a route for easing tissue stress associated with aging.

Rapamycin's effects extend into muscle as well. In this setting, it supports satellite cells—crucial contributors to muscle repair—by sustaining autophagy and reducing cellular strain. These effects are associated with changes in strength and mobility that are relevant to independence in later life. Rapamycin also influences metabolic processes. It alters the body's flexibility in shifting between glucose and fat use, while affecting insulin sensitivity and liver activity. Animal studies show changes in liver gene expression linked with adaptive metabolic responses under rapamycin exposure. Taken together, these findings place rapamycin among the most closely studied interventions in longevity research.

Revitalizing Immune Function and Protecting Cognitive Health

Immune systems naturally weaken with age, leaving older adults more vulnerable to infection and less responsive to vaccination. Rapamycin influences immune function in ways that may enhance this responsiveness. Clinical investigations have reported associations between rapamycin exposure, stronger vaccine responses, and fewer infections, supporting a role in maintaining immune performance later in life.

Rapamycin's effects extend into neural systems as well. Harmful proteins such as amyloid-beta and tau can accumulate in the aging brain, contributing to conditions such as Alzheimer's disease; Parkinson's disease involves abnormal alpha-synuclein accumulation. By activating autophagy—the cell's internal recycling system—rapamycin assists in clearing damaged proteins and preserving neural organization. In animal studies, rapamycin-treated mice have shown improvements in memory measures, reductions in neuroinflammation, and more intact neural circuitry. Ongoing human trials will determine how these immune-related and neurological findings translate into clinical relevance.

Insights from Animal Studies: Longevity Extended by Rapamycin

Consistent animal studies have linked rapamycin to lifespan extension across multiple species. In mouse models, treatment has been associated with increases in both median and maximum lifespan, including when administered later in life. Beyond effects on longevity, some studies report changes in age-related traits and selected physiological functions, highlighting the breadth of rapamycin's biological influence.

Extending these findings to humans requires caution, as larger, long-term human trials remain necessary before drawing firm conclusions about rapamycin's influence on healthspan or lifespan. Even so, lifespan-extension effects have been observed across a wide range of organisms—from yeast and worms to fruit flies—highlighting rapamycin's action on evolutionarily conserved biological pathways. This cross-species evidence keeps rapamycin at the forefront of contemporary longevity research.

Rapamycin in Human Research: Off-Label Use and Longevity Studies

The off-label use of rapamycin for longevity has drawn increasing interest from researchers and the public alike. Initially approved by the FDA to prevent kidney transplant rejection and manage specific medical conditions, rapamycin has since become a subject of investigation as a potential longevity intervention. Research protocols differ markedly from the high-dose, continuous regimens employed in transplant care, and instead favor intermittent, generally lower-intensity strategies. By periodically modulating activity within mTOR complex 1 (mTORC1)—rather than suppressing it continuously—these approaches engage cellular repair processes, moderate inflammatory signaling, and influence metabolic function without producing the degree of immune suppression seen in transplant protocols.

Research to date has outlined observations from experimental settings, primarily related to dosing frequency and exposure, informed by animal studies, clinical observations, and early human trials. These include intermittent and periodic dosing schedules and, in some exploratory models, low-intensity continuous regimens. All remain experimental in the context of aging and are not approved for longevity purposes. Investigators are also examining potential combination strategies, pairing rapamycin with agents such as metformin or acarbose. Findings from animal research, including work from the National Institute on Aging's Interventions Testing Program (ITP), suggest that some combinations influence aging-related pathways in complementary but distinct ways. The added complexity and interaction risks inherent in combination therapy require careful evaluation and do not constitute general guidance.

In clinical contexts where rapamycin is used for approved indications, treatment is initiated and monitored by a qualified clinician who considers each patient's history, comorbidities, and concurrent medications. Exploration of rapamycin for longevity purposes warrants individualized medical oversight and therefore remains investigational rather than established practice.

Safety, Risks, and Unanswered Questions

While rapamycin and the rapalogs represent significant developments in longevity research, their use requires balanced consideration of potential benefits, known risks, and unresolved uncertainties. Medical follow-up remains important to identify side effects early and guide thoughtful adjustments as treatment unfolds.

Across human studies and clinical reports, several adverse effects have been documented. Mouth ulcers are among the more commonly noted and are typically described as mild. Rapamycin and related compounds have also been associated with changes in lipid metabolism, including elevations in cholesterol or triglyceride levels. These observations highlight the importance of careful evaluation when interpreting findings from clinical use.

Because mTOR plays a central role in immune regulation, the timing and context of exposure are particularly important. Factors such as infection risk, planned surgeries, wound healing, and overall immune status are considered when therapy is initiated or adjusted. Early gastrointestinal symptoms—such as nausea, diarrhea, or mild abdominal discomfort—and occasional fatigue may occur at the start of treatment, typically settling as the body accommodates exposure or with modification.

Beyond these near-term considerations, longer-range biological questions remain. Although rapalogs often display more refined pharmacologic properties than rapamycin, uncertainties persist regarding long-term metabolic consequences. Sustained modulation of mTORC1 may, under certain conditions, influence mTORC2—a pathway involved in glucose regulation and insulin signaling—raising questions distinct from the lipid-related changes observed with some rapamycin regimens.

Future research must clarify how prolonged mTOR modulation affects tissues that rely heavily on this pathway, including skeletal muscle, connective tissue, and immune cells. Extended suppression of mTORC1 could influence muscle repair, wound-healing dynamics, or immune responsiveness. These possibilities underscore the importance of continued investigation rather than premature extrapolation.

Another unresolved issue is the potential for diminishing biological effects. Persistent mTORC1 inhibition may trigger adaptive responses that attenuate the intended influence of treatment. This has prompted researchers to examine strategies such as intermittent modulation, treatment cycling, or carefully structured combination approaches designed to preserve therapeutic consistency while limiting compensatory signaling.

Across research studies and reported clinical experience, attention has focused on the role of laboratory findings and overall clinical observations in evaluating mTOR-modulating therapies. Measures related to metabolism, liver function, blood parameters, and inflammation are commonly discussed in this context, alongside broader clinical status. Interest has also grown in how these therapies may relate to lifestyle factors such as nutrition, physical activity, and stress. While such interactions remain under study, they highlight the complexity of interpreting findings within diverse health contexts.

Although short-term human studies have produced encouraging findings, the long-term effects of intermittent mTOR modulation for aging remain uncertain. Current discussions of rapamycin and rapalogs therefore emphasize both their potential and their unresolved limitations.

Rapamycin's Path Forward: Future Human Trials and Research Directions

The future of rapamycin research continues to attract attention across longevity science and preventive medicine. Investigators are working to clarify how its biological effects differ across settings—how benefits and risks vary by context, and how different dosing strategies perform across study designs. The development of rapalogs—compounds related to rapamycin but engineered for distinct pharmacologic profiles—has grown out of this effort. Early findings suggest that these agents engage pathways involved in metabolic regulation, tissue-specific responses, and aging-related processes, though many questions remain unanswered.

One active area of investigation involves age-associated frailty, marked by gradual losses in strength, mobility, and physical resilience. Ongoing clinical studies are examining whether intermittent mTOR modulation, particularly when combined with resistance training, can help preserve muscle function and endurance. Results in this area would help clarify whether rapamycin or rapalogs can complement established strategies for maintaining physical capability later in life.

Immune function represents another central focus. Researchers are exploring whether mTOR activity can be modulated in ways that improve vaccine responsiveness and resistance to infection in older adults, without producing broad immune suppression. This line of work reflects a shift toward more targeted, age-specific approaches to immune regulation rather than uniform dampening of immune signaling.

Clarifying long-term safety and effectiveness through well-designed human trials remains essential. While animal studies have consistently demonstrated lifespan extension under specific conditions, human data are still limited. Current and forthcoming trials are assessing effects on cardiovascular health, metabolic regulation, neural protection, and outcomes such as frailty, functional decline, and recovery from illness or injury. Investigators are also studying combination strategies, including pairing mTOR modulators with senolytic agents, to determine whether distinct mechanisms produce additive or complementary effects. All of these approaches remain investigational and require careful evaluation.

As evidence continues to accumulate, rapamycin may eventually find a place within broader preventive or therapeutic frameworks. Approval for aging-related indications has not been established, and important questions about long-term safety remain. Ongoing research will determine whether mTOR-targeted interventions meaningfully influence aging biology while maintaining acceptable margins of risk.

The Rise of Rapalogs in Longevity Science

Building on advances made with rapamycin, researchers have developed rapalogs—modified compounds created to explore new uses and address some of rapamycin's practical limitations. Rapamycin's influence on pathways central to aging biology is well established in preclinical models and is now being explored in human studies. At the same time, clinical experience has revealed challenges, including immune suppression, variable pharmacokinetics, and complex dosing. Rapalogs arose from these constraints, incorporating structural refinements intended to improve targeting, bioavailability, and tolerability while preserving many of rapamycin's core biological effects.

Like rapamycin, rapalogs act on mTOR-related pathways, particularly mTORC1. Their structural differences, however, allow for more predictable pharmacologic behavior and, in some settings, distinct effects on immune function. Compounds such as everolimus—approved for cancer treatment and immune modulation—and ridaforolimus, which remains investigational, are being studied in aging biology, metabolic regulation, and immune responsiveness. Researchers are examining whether these agents influence metabolic processes, selectively shape immune activity, or interact with neural pathways relevant to aging in older adults. Through more sustained and better-characterized modulation of mTOR signaling, rapalogs are being explored as possible contributors to increasingly tailored approaches within longevity research.

Expanding the Potential of mTOR Modulation

Rapalogs share many of rapamycin's core biological actions—including the promotion of autophagy, effects on cellular metabolism, and modulation of inflammatory pathways—while incorporating structural changes intended to improve consistency and tolerability. This balance between targeted mTOR modulation and greater pharmacologic stability has made rapalogs a central focus of investigation, particularly in research that considers longer-term applications in aging biology.

Some rapalogs display more predictable pharmacokinetics than rapamycin, which in clinical settings often requires individualized dose adjustment. This has shifted attention toward whether these compounds can reduce some of the immune-related challenges observed with earlier regimens. Such considerations are especially important in aging research, where preserving immune competence while addressing cellular stress responses remains a central aim.

Individual rapalogs also differ in their tissue-specific characteristics, allowing them to be explored in distinct research contexts. Everolimus, for example, has been studied in older adults in trials examining vaccine responsiveness, contributing to broader investigations of immune aging. Rapalogs with shorter half-lives allow intermittent modulation of mTORC1 with defined recovery periods—an approach that echoes nutrient-responsive signaling patterns without maintaining continuous pathway suppression.

Beyond immune-focused research, rapalogs are being explored for their potential relevance to neural and metabolic function. In experimental models, mTORC1 modulation influences neuronal autophagy and selected neuroinflammatory processes and is associated with changes in pathways related to insulin sensitivity and lipid handling. Ongoing studies seek to clarify how the distinct properties of individual rapalogs might be applied in metabolic and neurological research, while avoiding complications linked to chronic or high-intensity rapamycin exposure.

Potential Synergies: Combining Rapalogs with Other Longevity Interventions

Experimental studies are examining combinations such as rapalogs with senolytics, including dasatinib and quercetin, to explore whether coordinated effects on senescent-cell accumulation and signaling create tissue environments more resilient to age-related stress.

Another line of research is evaluating combinations of rapalogs with caloric-restriction mimetics such as metformin or resveratrol—compounds that activate pathways like AMPK and sirtuins and converge with mTOR-related mechanisms. Investigators are assessing whether these pairings affect metabolic processes in ways that align with intermittent mTOR modulation, particularly in contexts where sustained pathway suppression is undesirable. NAD^+ precursors such as NMN or NR are also being examined for roles in mitochondrial function and cellular repair pathways, including when evaluated alongside rapalogs.

All of these strategies remain experimental and are evaluated in controlled research settings. They are not established therapies and should not be regarded as general clinical guidance.

Rapamycin, Rapalogs, and the Dawn of Age-Modifying Medicine

Rapamycin and its derivatives, the rapalogs, have become important areas of investigation within longevity research. Originally developed as immunosuppressants, these compounds now attract scientific interest for their influence on pathways central to aging biology in experimental models. Preclinical studies suggest effects on muscle integrity, metabolic regulation, chronic inflammation, and select aspects of immune and neural function. While animal studies consistently demonstrate lifespan-related outcomes under specific conditions, these findings justify continued investigation rather than predict benefit in humans.

Many questions remain unresolved. Extensive clinical research continues to clarify safety profiles, optimal dosing strategies, and long-term effects in humans. Investigators are examining how variables such as age, comorbidities, concurrent medications, and cumulative exposure shape both potential benefits and risks. As with any agent that acts on central regulatory pathways, careful assessment of long-term safety—including metabolic, immune, and tissue-specific responses—remains essential.

THE ART, SCIENCE, AND STRATEGY OF LONGEVITY

As longevity science moves toward greater biological specificity, rapamycin and the rapalogs may contribute to research-driven strategies that examine variation across genetic background, baseline health, and biomarker responses. Such work may help clarify where these compounds show promise and where limits remain, supporting more informed interpretation of their potential roles.

These scientific questions unfold alongside broader reflections about aging. If future research reveals ways to sustain capability or delay aspects of functional decline, how might that influence the goals people set, the ambitions they pursue, or the meanings they attach to different stages of life? What ethical principles should govern fair access to interventions that act on aging biology? And is society—economically, socially, and philosophically—prepared for a future in which aspects of aging can be altered while the fact of aging remains unchanged?

CHAPTER 15

Metformin: Reimagining a Diabetes Drug for Longevity

"Discovery consists of seeing what everybody has seen and thinking what nobody has thought."
—ALBERT SZENT-GYORGI

For decades, metformin has served as a first-line medication for type 2 diabetes, used for its role in regulating blood glucose. More recently, this familiar drug has drawn attention for a different reason: its relevance to biological processes associated with aging. Metformin influences cellular stress responses, metabolic regulation, and inflammatory signaling. These findings have prompted investigation into whether its broader physiological effects intersect with aging biology, even as questions about human longevity remain unresolved.

Unlike many experimental interventions, metformin is widely available, affordable, and extensively studied in humans for established medical indications. Its effects extend beyond glucose control, involving metabolism, cellular maintenance, inflammatory pathways, and gene expression. Rather than acting as a single-purpose drug, it influences several domains of physiology, with research examining how these actions relate to aging biology. Even so, most current evidence comes from observational analyses and mechanistic studies rather than randomized trials designed to evaluate aging outcomes directly. This leaves an essential question open: do associations with healthier aging reflect direct engagement with aging biology, or are they largely a consequence of improved metabolic function?

Unlocking Longevity Pathways: How Metformin Works

At the cellular level, aging reflects an interplay of metabolic shifts, chronic inflammation, and the slow accumulation of molecular injury. Metformin interacts with several of these processes at once. One of its most studied effects involves activation of AMP-activated protein kinase (AMPK), a central regulator of cellular energy balance. When AMPK is stimulated, metformin has been linked to changes in mitochondrial activity, glucose utilization, and insulin signaling—features that echo metabolic adaptations observed in caloric-restriction studies in animal models.

Metformin's reach extends beyond energy metabolism. Its effects on the mechanistic Target of Rapamycin (mTOR) pathway—a key coordinator of cellular growth and repair—have also been examined. Modulating mTOR activity can permit periods of internal restoration, and in laboratory models this shift is associated with enhanced autophagy, the process through which cells clear damaged proteins and worn cellular components. Autophagy supports cellular stability by limiting the buildup of molecular debris, a feature often observed in age-related physiological decline.

Aging is also marked by persistent, low-grade inflammatory activity. Metformin has been associated in experimental and clinical studies with changes in inflammatory pathways such as NF-κB. Its relationship with alterations in the gut microbiome—a community that contributes to metabolic and immune regulation—has also been explored.

Emerging evidence points to another possibility: metformin may interact with epigenetic mechanisms, including patterns of DNA methylation that influence gene expression across the lifespan. Although this area remains early, these findings suggest a link between metformin and molecular markers used to study biological aging.

Viewed together, these mechanisms describe a drug that intersects with several domains of aging biology—from metabolism and inflammation to cellular maintenance and possible epigenetic regulation. Yet the essential question endures: will these cellular and molecular signatures correspond to measurable changes in human aging outcomes? Resolving that question depends on long-term, rigorously controlled trials.

Insights from Animal and Human Studies

Across a wide range of species—from simple organisms to mammals—metformin has been linked to extended lifespans and delayed features of age-related decline. In mice, for example, metformin-treated animals exhibit distinct metabolic signatures and show differences in the incidence of conditions such as cancer, cardiovascular dysfunction, and cognitive impairment. Many maintain strength and mobility into advanced age, offering a clear view of aging-related traits in preclinical models.

Studies in *Caenorhabditis elegans* (*C. elegans*), the microscopic worm widely used in aging research, reveal similar findings. In these organisms, metformin alters gut-microbiome composition and modifies metabolic pathways associated with lifespan in experimental settings. These observations underscore the relationship between microbial communities and longevity, a connection now being examined in more complex animals.

Research involving nonhuman primates has identified metformin-associated effects across multiple organ systems, including the kidneys, lungs, skin, and brain. Of particular interest are findings showing reduced markers of cellular senescence, including lower levels of senescent-cell accumulation—cells that persist in tissues and release inflammatory compounds. In these models, metformin is associated with indicators of tissue stability and more gradual physiological change with advancing age.

Human evidence, primarily observational, offers consistent associations. Across large population studies, people with type 2 diabetes who use metformin often demonstrate lower recorded mortality rates compared with those using alternative glucose-lowering medications. This unexpected pattern—sometimes referred to as the "metformin paradox"—has contributed to sustained scientific interest in the drug's broader biological effects. Some analyses report that metformin users with diabetes display outcomes that approach those of non-diabetic counterparts. These findings are correlational and require careful interpretation.

Whether metformin directly influences human aging depends on controlled trials designed for that purpose. The anticipated TAME Trial (Targeting Aging with Metformin) represents one such effort, aiming to assess whether metformin can alter the trajectory of age-related conditions in a structured clinical environment. Studies of this kind are needed to determine whether observations from animal models and epidemiologic analyses translate into measurable changes in human aging.

The TAME Trial: A Defining Test for Metformin's Longevity Potential

Aging has long been viewed as an inevitable decline, and research now examines whether certain biological features of aging can be modulated. Investigating this question is the central aim of the Targeting Aging with Metformin (TAME) Trial. Unlike traditional clinical studies that concentrate on single diseases, TAME is structured to examine whether metformin influences the timing or progression of multiple age-related conditions in parallel. If the trial yields clear results, it could help shift aging from a background assumption in medicine to a set of measurable biological processes with **defined** endpoints.

The implications of TAME extend beyond evaluating metformin itself. Should the trial provide evidence that a pharmacological agent affects several aging-related outcomes together, it would offer a rationale for regulators, including the FDA, to consider how therapies directed at aging biology are evaluated. This type of evidence could support the development of interventions aimed at maintaining functional capacity later in life and encourage broader investigation across therapeutic fields.

The challenges of a trial like TAME remain considerable. Aging is not classified as a disease, which complicates the design and approval of studies that directly examine aging biology. Furthermore, human aging reflects contributions from genetic, metabolic, environmental, and behavioral factors, making it difficult to isolate the effects of a single intervention. Large trials require extended follow-up, substantial resources, and diverse study populations to produce findings that are robust and widely applicable.

If TAME succeeds despite these constraints, the outcome may reshape future research priorities. Demonstrating whether an established, widely used medication can alter biological correlates associated with age-related decline would strengthen the case for studying aging as a coordinated set of processes rather than a series of isolated conditions. Such findings could guide development of therapies ranging from agents that activate AMPK or modulate mTOR to approaches designed to address senescent cells or support mitochondrial function.

As the field continues to advance, the TAME Trial stands as a critical waypoint in efforts to understand the mechanisms that accompany aging. Whether metformin becomes one component of a broader set of strategies or serves primarily to clarify how aging should be studied, its findings will guide how aging biology is studied and how next-generation interventions are developed.

Metformin and Exercise: Balancing Metabolic Benefits and Physical Adaptation

The relationship between metformin and exercise presents a real paradox. Each has been shown to affect metabolic function, insulin sensitivity, and mitochondrial activity, yet their combined influences do not always align in consistent or predictable ways. Metformin, widely studied for its metabolic actions in clinical and research contexts, intersects with several of the same cellular pathways engaged by exercise, introducing the possibility of competing or overlapping effects. How these influences converge varies among people, shaped by physiology, training goals, and baseline metabolic characteristics. Clarifying this interaction helps place metformin in the context of an active lifestyle.

Exercise produces well-documented adaptations in metabolic and mitochondrial function, including improved insulin sensitivity, enhanced mitochondrial efficiency, and greater capacity to manage oxidative stress. Metformin engages many of these same systems in experimental settings, but their interaction is not uniform. In some studies, metformin parallels exercise-related metabolic responses; in others, it alters pathways involved in endurance, muscle adaptation, or energy use. Together, these findings underscore that metformin and exercise act within shared physiological networks, but do not necessarily reinforce one another in predictable ways.

Metformin's Influence on Exercise Adaptations

Metformin influences exercise adaptations through mechanisms that intersect with—but do not replicate—those activated by physical activity. The drug affects several of the same cellular pathways involved in exercise, including AMP-activated protein kinase (AMPK), a sensor that regulates glucose uptake, fatty-acid use, and mitochondrial function. AMPK activation, however, also has downstream consequences: it can reduce signaling through the mechanistic Target of Rapamycin (mTOR), a pathway central to muscle protein synthesis. This combination of activating AMPK while dampening mTOR has prompted discussion about how metformin interacts with resistance-training goals.

Mitochondrial dynamics introduce further complexity. Metformin inhibits mitochondrial complex I, an enzyme essential for ATP production during sustained or high-intensity activity. Because ATP availability influences muscular performance, complex I inhibition can affect energy output under certain conditions. Many people do not experience measurable changes, but individuals engaged in high-volume endurance work or intensive athletic training may show greater sensitivity to such effects. This overlap helps explain why metformin and exercise act on the same molecular systems yet do not always produce parallel physiological outcomes.

Mixed Outcomes: When Metformin and Exercise Collide

The combined effects of metformin and exercise vary considerably, influenced by exercise type, intensity, duration, and individual metabolic characteristics. Some studies report that metformin modifies the degree of improvement in insulin sensitivity typically observed after exercise training. These findings indicate that pairing two interventions known for metabolic effects does not necessarily produce additive outcomes, and that their interaction differs across contexts.

Research has also examined whether metformin's activation of AMPK and its influence on mTOR signaling affect pathways involved in strength or hypertrophic adaptations from resistance training. Findings are mixed. Some studies describe modest reductions in muscle growth, while others report minimal or no measurable differences compared with exercise alone. Taken together, these results point to variability in physiological response shaped by training demands, baseline health, and metabolic context rather than a single predictable interaction.

Who Benefits Most? Individual Considerations

Metformin's influence on exercise adaptations is not uniform; it varies with age, metabolic status, and individual objectives. In older adults, pathways involved in strength and muscle maintenance draw particular attention because metformin engages mechanisms linked to muscle growth and repair. This has prompted investigation into how metformin relates to resistance-training outcomes in people without metabolic conditions who focus on improving muscular adaptation. For those managing insulin resistance or other metabolic disturbances, metformin's established role in glucose regulation—and its documented links with inflammatory pathways—often carries greater relevance than any potential influence on strength or hypertrophy. Research continues to explore how physical training and metformin interact across these scenarios.

Performance-focused athletes face a different research landscape. High-level training relies heavily on mTOR signaling and mitochondrial function, both of which metformin influences. Investigators have assessed whether these effects intersect with endurance- or strength-oriented adaptations, an area that remains under active study for those whose training depends on maximizing physical output.

For people who exercise primarily for general metabolic support or long-term physiological stability, these mechanisms are less central. Activities such as walking, cycling, and moderate resistance work are less influenced by the pathways most often examined in metformin research. In these settings, studies of metformin's metabolic interactions help clarify how the drug relates to routine physical activity.

Metformin and Longevity: Final Reflections

Originally developed as a diabetes medication, metformin has become a central feature of scientific inquiry into aging biology. Research examining metabolic pathways, inflammatory signaling, mitochondrial activity, and cellular maintenance has placed it in a distinctive position within this field. More than a tool for glucose regulation, it has served as a means of probing how interconnected biological processes contribute to aging and how they can be studied systematically.

Few medications combine metformin's affordability, extensive human safety data in approved uses, and wide range of mechanistic links. These characteristics have made it a recurring focus of aging-related research, even as responses to the medication vary. Metabolic status, genetics, physical activity, and other physiological factors influence how metformin interacts with the body.

For those managing insulin resistance or metabolic disturbances, metformin's established clinical role remains relevant. Adults without metabolic conditions and with goals related to muscle growth or athletic performance often consider its role differently, particularly given investigations into how it engages pathways involved in hypertrophy and endurance adaptations. These contrasts reflect the diversity of physiological states and research questions rather than any single pattern of response.

Ongoing scientific efforts, including large-scale clinical trials, seek to clarify where metformin fits within the broader landscape of aging research. Whether it becomes a central component of future investigation or primarily helps define the path toward more specialized interventions, its role in shaping current scientific understanding is evident. The broader hope lies in the continuing refinement of aging biology—an effort to which metformin has contributed meaningfully. As research advances, the field moves toward a clearer view of the processes that underlie aging and toward a deeper understanding of how those processes can be examined with increasing precision.

CHAPTER 16

NAD⁺ and Aging: NMN, NR, and Cellular Energy

"Progress is not in enhancing what is, but in advancing toward what will be."
—KHALIL GIBRAN

NAD$^+$ is a universal coenzyme found in all cells and involved in the fundamental reactions that sustain life. It participates in energy metabolism, genomic maintenance, and regulatory processes that enable cells to respond to physiological demands. Because so many systems rely on NAD$^+$, changes in its availability directly influence cellular function across the lifespan. As NAD$^+$ levels decline with age, these interconnected processes operate under increasing strain—an observation that has made NAD$^+$ a central focus in research examining the biology of aging.

Energy metabolism is one of NAD$^+$'s most visible roles. By transferring electrons to the mitochondrial electron transport chain, NADH (the reduced form of NAD$^+$) supports the production of ATP, the primary source of usable energy. This contribution allows cells to meet routine metabolic needs and adapt to fluctuations in energy demand. When NAD$^+$ becomes limited, mitochondrial function adjusts in ways consistent with constrained energy availability.

Genomic maintenance also depends on NAD^+. DNA is continually subject to metabolic and environmental sources of damage, and repair enzymes—including poly(ADP-ribose) polymerases (PARPs)—require NAD^+ to preserve genomic integrity. PARP activity increases when repair demands rise, and this increased activity consumes NAD^+. When NAD^+ availability is low, repair capacity becomes more difficult to sustain, creating tension between the need for genomic stability and the resources available to support it.

NAD^+ also supports the activity of the sirtuin family of proteins—SIRT1, SIRT3, and others—which influence metabolic processes, inflammatory signaling, and cellular adaptation to stress. SIRT1 participates in coordinating energy expenditure and gene expression, while SIRT3 contributes to mitochondrial homeostasis and defense against metabolic stress. Because sirtuins use NAD^+ as a cofactor, fluctuations in NAD^+ levels directly shape how these systems operate, particularly under conditions that challenge cellular resilience.

These systems interact in numerous ways. Oxidative and metabolic stress can increase PARP activation, consuming NAD^+. Reduced NAD^+ can limit sirtuin activity, altering stress responses and mitochondrial function. Mitochondrial adjustments, in turn, influence the balance between NAD^+ and NADH, affecting redox status in ways that ripple across energy pathways. Taken together, these connections place NAD^+ at a convergence point where metabolic, genomic, and stress-response systems meet, each drawing from a shared pool.

Age-Related Decline and Regulation of NAD^+

Research consistently shows that NAD^+ levels decrease with age in many tissues. This decline does not stem from a single pathway but reflects changes in production, recycling, consumption, and metabolic demand. The result is a gradual shift in how cells sustain processes that rely heavily on NAD^+.

Much of the NAD^+ used in adult tissues comes from the salvage pathway, which regenerates NAD^+ from nicotinamide using the enzyme

NAMPT (nicotinamide phosphoribosyltransferase). NAMPT activity declines with age in several experimental models. Because this pathway supports most NAD$^+$ production, even modest changes in NAMPT function can influence cellular NAD$^+$ levels. Interest in the salvage pathway has grown as researchers seek to understand how cells maintain NAD$^+$ under changing physiological conditions.

Another factor shaping NAD$^+$ availability is consumption by enzymes involved in immune signaling and cellular communication. CD38, an enzyme often found on immune cells, is a significant consumer of NAD$^+$. CD38 expression increases with age, and inflammatory states further amplify this increase. Because CD38 breaks down NAD$^+$ during its activity, elevated levels accelerate NAD$^+$ depletion. These observations have prompted closer examination of how immune activation, inflammation, and metabolic stress interact with NAD$^+$ regulation.

Redox balance provides additional context. The ratio of NAD$^+$ to NADH influences mitochondrial function, metabolic pathways, and cellular signaling. Conditions such as metabolic overload, alcohol metabolism, or disrupted nutrient processing shift this ratio, altering redox status in ways that affect NAD$^+$ availability. These shifts also influence how cells allocate resources between metabolic activity and repair, creating further feedback between energy systems and NAD$^+$ use.

Circadian rhythms play a role as well. Enzymes involved in NAD$^+$ production and consumption follow daily cycles aligned with sleep, feeding patterns, and metabolic activity. When circadian rhythms become irregular—due to aging, lifestyle, or environmental change—these cycles are disrupted, influencing NAD$^+$ metabolism. Research continues to examine how circadian regulation intersects with broader themes in aging biology.

Diet, activity, and energy balance shape NAD$^+$ availability by altering pathways that rely on this coenzyme. Physical activity affects enzymes that regulate both NAD$^+$ production and use. Nutrient composition influences metabolic flux, and caloric restriction or fasting alters how cells distribute NAD$^+$ across competing processes. Alcohol metabolism consumes NAD$^+$

in reactions that modify cellular redox states. Together, these relationships illustrate how common physiological states interact with NAD⁺ biology without implying specific outcomes or clinical conclusions.

NAD⁺ Precursors: NMN and NR in Metabolic Pathways

Because NAD⁺ crosses cell membranes inefficiently, scientific attention has turned to precursor molecules that enter biosynthetic pathways more readily. Nicotinamide mononucleotide (NMN) and nicotinamide riboside (NR) are two such compounds that feed into the salvage pathway.

NMN enters NAD⁺ biosynthesis through a direct enzymatic step, and many tissues contain the machinery necessary to convert NMN efficiently. Although NMN occurs naturally in foods, dietary levels remain far below those used in experimental research. Studies continue to examine how NMN availability, metabolic conditions, and tissue-specific factors influence NAD⁺ levels.

NR follows a related route but is converted into NMN before entering NAD⁺ synthesis. Present in small amounts in milk and yeast, NR has been studied for its tissue distribution, particularly in organs involved in metabolic regulation and neuronal activity. Whether NR itself crosses protective barriers around the brain, and how this relates to cellular energy processes, remains under investigation.

NMN and NR engage pathways involved in DNA repair, mitochondrial function, and NAD⁺-dependent enzyme activity. Research explores how these molecules influence mitochondrial metabolism, sirtuin activity, and cellular responses to metabolic stress. NMN has often been examined in relation to muscular and vascular energy demands, while NR has been studied for its connections to neuronal maintenance and stress-response pathways.

Animal studies have raised questions about metabolic adaptation and longevity, prompting interest in whether precursor supplementation alters NAD⁺ dynamics in experimental settings. Translating these findings to humans remains complex. Human studies are still limited, and definitive conclusions have yet to emerge. Ongoing work aims to clarify how NMN and NR interact with NAD⁺ pathways under differing physiological conditions.

Although interest in NAD⁺ biology continues to expand, questions remain about how NMN and NR function during age-related changes in metabolism and cellular regulation. These compounds take part in pathways that help cells make energy, manage oxidation, and respond to stress, but their impact varies depending on the tissue involved, baseline physiology, and how the body controls NAD⁺ turnover. Because they interact with enzymes that both build and use NAD⁺, their influence reflects the balance between these processes rather than a simple rise in NAD⁺ levels.

Human research remains early, and current studies differ in duration, formulations, and measured outcomes. Some show changes in markers related to energy use or metabolic activity, while others find smaller or inconsistent shifts. Researchers are examining how factors such as precursor availability, daily habits, and genetics contribute to this range of outcomes. Ongoing work is expected to clarify which tissues—such as muscle, liver, or blood vessels—and which physiological states respond more readily to changes in precursor intake.

These findings point to the value of considering NMN and NR alongside other established influences on aging biology. Movement increases mitochondrial demand, nutrition affects substrate availability, and sleep contributes to circadian regulation of metabolic pathways. When NAD⁺ precursors are viewed within this wider biological setting, their place in aging-related research becomes easier to understand than when they are approached as isolated interventions.

Conclusion: NAD⁺ in Aging Biology

NAD⁺ occupies a distinctive position in aging biology because it sits at the intersection of energy production, genomic repair, and cellular regulation. Rather than acting within a single pathway, it moves continuously between systems that sustain metabolism, preserve DNA integrity, and coordinate responses to physiological stress. Its availability reflects the interplay between processes that replenish NAD^+ and those that consume it—an equilibrium that shifts with age, metabolic state, immune activity, and environmental demands.

Research on NAD^+ decline has drawn attention to how tightly coupled these systems are. Changes in NAD^+ metabolism do not occur in isolation; they emerge from interactions among mitochondrial function, inflammatory signaling, redox balance, circadian rhythms, and cellular repair mechanisms. Interest in NMN, NR, and related compounds arises from efforts to understand these interactions more clearly—not as isolated solutions, but as tools for probing how aging alters resource allocation within cells.

Collectively, studies of NAD^+ biology reinforce a broader theme in aging research: aging reflects coordinated changes across interconnected systems rather than failure of a single process. The value of this work lies less in any one intervention and more in the insight it provides into how cells sustain essential functions over time, adapt to cumulative stress, and manage competing demands as physiological conditions change. NAD^+ research thus contributes to a deeper understanding of aging as a dynamic biological process, shaped by constraint, regulation, and adaptation rather than simple decline.

CHAPTER 17

Targeting Aging at the Cellular Level: Senolytics, Metabolism, and Telomeres

> *"For age is opportunity no less than youth itself, though in another dress, and as the evening twilight fades away, the sky is filled with stars, invisible by day."*
> —HENRY WADSWORTH LONGFELLOW

A major focus of aging science now centers on cellular senescence. Senescent cells stop dividing but remain metabolically active in the body, and they appear throughout life in response to injury, cellular damage, or genomic instability. Early in life, they support wound repair and help contain unstable cells. With age, senescent cells accumulate in tissues, where their presence is associated with changes in tissue structure, increased inflammatory signaling, and altered responses in neighboring cells. Studying these shifts helps explain how tissues adapt to ongoing damage and physiological strain.

Interest in these processes has led to growing research on senolytics, a group of compounds examined for their effects on pathways associated with senescent cells. Current studies focus on how these compounds influence the chemical signals released by senescent cells and how those signals relate to changes in tissue behavior. Most findings remain preclinical, but they have expanded understanding of how cellular defense and response systems operate during aging.

Senescence research also intersects with other areas of cellular biology. Telomerase helps protect chromosome ends and supports genomic stability. Mitochondria regulate energy production and contribute to the generation of reactive oxygen species. Metabolic signaling guides how cells respond to nutrient availability and cellular demand. Together, these systems describe how cells maintain function over time without reference to specific health outcomes.

This chapter reviews the biology of senescent cells, emerging work on senolytic compounds, and the roles of telomerase, mitochondrial function, and metabolic regulation in aging research. Considered together, these areas illustrate how cells manage persistent demands and adapt to the pressures that accumulate across the lifespan.

The Dual Nature of Cellular Senescence

Cellular senescence holds an important place in aging biology. When cells face substantial stress—such as oxidative injury, DNA damage, or repeated cycles of division—they may enter a senescent state. This response halts further replication and limits the spread of compromised material. Early in life, it supports tissue protection and damage control. With age, however, senescent cells tend to persist in tissues for longer periods. Their persistence has drawn interest because it alters the behavior of nearby cells and contributes to broader biological changes observed later in life.

A defining feature of senescent cells is the senescence-associated secretory phenotype (SASP), a collection of molecules released into their surroundings. These signals modify the extracellular environment and activate stress responses in neighboring tissue. The chronic, low-grade inflammatory activity associated with SASP—often called "inflammaging"—reflects ongoing shifts in tissue signaling rather than a discrete pathological condition.

Senescent cells also send signals that can prompt nearby cells to adopt similar stress-response states, a process sometimes referred to as secondary senescence. At the same time, the immune system's ability to recognize and clear senescent cells changes with age. As senescent cells accumulate, they influence tissue structure, cellular turnover, and metabolic regulation across broader regions.

Senolytics: Investigating Compounds That Target Senescent-Cell Pathways

Against this backdrop, senolytics have become a focus of research into compounds that interact with pathways associated with senescent cells. These agents are studied for their ability to selectively eliminate senescent cells and to alter the signaling features that define senescence. Researchers also investigate how changes in these signals relate to shifts in nearby cell behavior and tissue structure. Most available evidence comes from preclinical models, which allow detailed examination of senescent-cell dynamics without extending conclusions to human biology.

Because senescent cells release a distinct set of molecules through the SASP, senolytic research often asks whether altering senescent-cell burden changes these molecular signals. Preclinical studies have reported associations between senolytic exposure and shifts in gene activity, inflammatory signaling, and other markers that vary with age. These findings frame the questions that guide ongoing investigation, while leaving implications for human physiology unresolved.

Senescent cells also play important roles in wound repair, tissue remodeling, and other adaptive responses. For this reason, eliminating all senescent cells is neither feasible nor desirable. This complexity has led researchers to explore whether modulating senescent-cell presence in specific contexts—rather than pursuing complete removal—offers a clearer view of how these cells influence tissue behavior as biological demands change.

Advancing Approaches to Senescent-Cell Biology

The growing field of senolytic research reflects a broader effort to understand how senescent cells communicate with their environment and which pathways support their persistence. Early work focused on only a few compounds, but the field now encompasses a wider range of pharmaceutical agents and naturally occurring molecules. Each interacts with senescent cells through distinct mechanisms, highlighting the complexity of senescence and the varied strategies used to study it.

Navitoclax (ABT-263), originally developed for cancer treatment, is one example of this expanded work. It acts on BCL-2 family proteins, which help determine whether a cell survives under stress. In research settings, navitoclax has been used to examine how these survival pathways relate to senescence. Its known effects on platelets and other blood cells also illustrate important considerations when studying compounds that act on these systems.

FOXO4-DRI represents a different line of inquiry in senolytic research. It was designed to disrupt the interaction between FOXO4 and p53, proteins that regulate cellular stress responses. In preclinical models, FOXO4-DRI is used to probe how senescent cells maintain their state and which signals contribute to that persistence. These investigations address specific mechanistic questions without extending conclusions to human biology.

Naturally occurring molecules offer additional avenues for investigation. Apigenin, found in parsley, chamomile, and celery, has been studied for its effects on pathways related to inflammation and oxidative stress. EGCG, a compound in green tea, has been examined in similar laboratory contexts. Such studies allow researchers to observe stress-response pathways under controlled conditions. This work remains preclinical, and its relevance to humans depends on further investigation.

Taken together, these lines of inquiry deepen understanding of senescence and illustrate how cells manage persistent stress. Emerging senolytic candidates do not yield definitive conclusions at this stage. Instead, they sharpen the questions guiding ongoing work and situate senescence within the broader study of aging biology. As the field develops, its primary value lies in clarifying underlying mechanisms rather than anticipating near-term clinical application.

Telomerase Pathways in Cellular Aging

In contemporary aging research, senolytic studies and telomerase investigations offer complementary perspectives on how cells change over time. Senolytic work examines pathways linked to the accumulation of senescent cells, while telomerase research focuses on how cells maintain telomeres—the protective DNA segments at chromosome ends that shorten with each round of division. Because telomere length reflects a cell's history of replication, telomerase plays a central role in understanding how cells manage long-term demands on renewal.

Telomerase activity is often examined in tissues that undergo continuous turnover, including components of the immune system, the skin, and the gastrointestinal lining. These tissues rely on repeated cycles of regeneration, and telomere maintenance influences how they respond to ongoing replication and environmental stress. Attention has increasingly turned to how telomerase function intersects with other systems that regulate these responses.

Telomerase also presents interpretive challenges. The enzyme supports telomere preservation, yet dysregulated telomerase activity is associated with several cancers. This dual role requires that telomerase be considered within a broader biological context: it contributes to genomic stability while also intersecting with pathways involved in uncontrolled cell growth.

Viewed alongside senolytic research, telomerase studies help frame aging as a set of interacting cellular processes. One line of inquiry examines how tissues respond as senescent cells accumulate; another considers how cells maintain the structures required for continued division. Together, these approaches describe how cells adapt to ongoing demands across the lifespan, revealing both the limits of replication and the mechanisms that sustain cellular function.

Metabolic Regulation and Acarbose in Aging Biology

Metabolic regulation plays an important role in aging research, and acarbose has drawn attention for its effects on several related processes.

Although most familiar in glucose-management settings, researchers are now examining how its influence on carbohydrate digestion relates to cellular conditions that evolve over time.

Acarbose inhibits alpha-glucosidase in the small intestine, slowing the breakdown of starches and sugars. This change has prompted interest in how it affects metabolites formed during digestion, as well as glycation-related products such as advanced glycation end-products (AGEs). These molecules can alter tissue structure and modify cellular signaling. Current studies use this connection to explore how long-term digestive activity corresponds with physiological changes observed with age.

Acarbose has also been studied in relation to gut-microbiome activity. Because some of the compound passes into the large intestine unabsorbed, it becomes available for microbial fermentation. This has focused attention on its relationship to the production of short-chain fatty acids such as butyrate, which participate in communication between microbial populations and host tissues. These interactions help clarify how metabolic regulation responds to internal pressures and environmental exposures.

Animal studies have reported associations between acarbose exposure and changes in metabolic or inflammatory markers. These findings inform ongoing questions about how metabolic processes adjust with age and how those adjustments relate to broader features of aging biology.

Cellular Aging as an Interacting System

Research on senolytics, telomerase activity, and metabolic regulation helps define the cellular changes that accompany aging over time. Each area addresses a different dimension of this biology. Senolytic studies examine how senescent cells persist, communicate with surrounding tissue, and influence local stress responses. Telomerase research focuses on how cells maintain chromosome integrity across repeated cycles of division. Metabolic investigations describe how energy handling, mitochondrial function, and nutrient signaling adjust as physiological demands accumulate. Taken together, these fields outline the mechanisms that support cellular maintenance, repair, and adaptation.

Viewed in combination, these mechanisms reveal aging as a process shaped by interaction rather than isolation. Cells do not age because a single pathway fails, but because multiple regulatory processes adjust simultaneously under sustained pressure. Senescence limits the propagation of damaged cells while altering tissue signaling. Telomere maintenance constrains replicative capacity while preserving genomic stability. Metabolic regulation allocates energy and resources in ways that balance immediate demands against long-term function. Each imposes boundaries, yet each also contributes to cellular resilience.

Much of this work remains in development, particularly in human studies, but examining these mechanisms together clarifies how aging unfolds at the cellular level. Aging emerges not as a uniform decline, but as the cumulative result of trade-offs between repair, renewal, and resource management. The value of integrating senescence, telomerase biology, and metabolic regulation lies in revealing aging as a coordinated biological condition—defined by constraint, adaptation, and persistent stress—rather than as a collection of independent cellular failures.

CHAPTER 18

Non-Pharmacological Biohacking: Tools and Approaches Beyond Medication

"The only way to discover the limits of the possible is to venture a little way past them into the impossible."
—ARTHUR C. CLARKE

Longevity science has long focused on pharmaceuticals and supplements, but a substantial share of biological regulation arises from signals the body receives from its surroundings. Non-pharmacological biohacking examines these signals—light exposure, temperature variation, movement, and other physical forces—to understand how the body adapts without relying on drugs. Simple tools and environmental adjustments can be used to track shifts in heart rate, energy use, alertness, and temperature regulation during controlled exposures. These approaches are not treatments; they offer ways to examine how daily routines and surroundings relate to processes involved in aging.

Interest in this area has expanded because it draws attention to aspects of aging often missed in standard clinical evaluation. Pharmaceuticals tend to act on specific molecular pathways, whereas environmental and behavioral inputs engage broader physiological responses to light, temperature, mechanical load, and activity demands.

These exposures have been examined in relation to measurable shifts in cellular activity, mitochondrial function, autonomic signaling, and the mental and physical effort sustained across the day. Evidence varies, but taken together, these findings clarify how physiology responds to environmental and mechanical forces.

Individual responses depend on several factors. Medical history, sensitivity to temperature or light, and general health influence how the nervous system and other systems register and interpret these inputs. Some responses are well documented, while others remain under study, making it important to view these observations within a broader health context rather than assume uniform effects.

Many such exposures occur naturally in daily life. Morning light during a walk, brief cold exposure at the end of a shower, or time spent in warmth after exercise are common conditions under which the body adjusts to changing demands. These situations provide practical contexts for observing how specific environmental forces correspond with physiological responses.

Non-pharmacological biohacking now appears alongside pharmacologic research within a broader effort to understand how lifestyle, environment, and technology interact with aging. These approaches generate observations rather than predictions and can be considered alongside clinical findings. They may reveal changes in autonomic activity, energy expenditure, sleep—wake behavior, or cognitive performance when the body encounters shifts in temperature, light, mechanical load, or controlled breathing. Ongoing work continues to examine how these responses relate to cellular- and tissue-level processes involved in aging.

This chapter examines the major non-pharmacological approaches used to observe how the body adjusts to environmental and physical inputs, and how these observations fit within current research on aging biology.

Real-Time Tracking and Physiological Measurement

Real-time tracking offers a more detailed view of day-to-day physiology than intermittent measurements alone. Devices that operate continuously or at frequent intervals reveal how glucose levels, heart-rate variability, sleep stages, body temperature, and markers of sympathetic activity shift across the day. These tools do not replace clinical evaluation; they add context by capturing short- and longer-term physiological variation.

Wearable devices—including watches, rings, and fitness trackers—collect data on movement, rest, heart-rate variability, and activity patterns. Accuracy varies with sensor design, skin characteristics, and external factors such as temperature or motion. Clinical assessment and periodic laboratory testing provide context, helping distinguish consistent physiological signals from device-related noise.

Continuous glucose monitors (CGMs) generate a steady stream of glucose measurements, illustrating how meals, physical activity, stress, and sleep relate to changes over time. Research has linked certain glucose profiles with metabolic features such as insulin sensitivity or variability, though interpretation remains highly individual and dependent on timing, conditions, and associated symptoms.

Tools that assess energy use—such as devices measuring resting metabolic rate or respiratory exchange ratio (RER)—indicate how the body shifts among fuel sources under different circumstances. Athletes and others focused on physical workload sometimes track these measures to better understand metabolic demands, but responses vary widely and depend on testing conditions and recent activity.

Hormone assessments obtained through saliva, blood, or urine offer insight into changes in hormones involved in mood, energy regulation, and reproductive and metabolic function. Age-related variation in hormones such as testosterone and estrogen has been examined in relation to shifts in energy levels, body composition, and other physiological measures. In clinical settings, these assessments contribute to broader evaluations of how hormone levels change with aging.

Sleep-tracking technologies—including EEG headbands, smart mattresses, and motion sensors—capture sleep stages, interruptions, and movement throughout the night. These recordings may reflect disrupted sleep, environmental influences such as light or temperature, or irregular schedules, and can help identify patterns that warrant closer attention.

The value of real-time tracking lies in the measurements, not the devices themselves. These tools make physiological change easier to observe and help connect daily actions with measurable responses. When interpreted carefully and paired with clinical judgment, real-time tracking supports a clearer understanding of how the body responds to meals, activity, stress, sleep, temperature shifts, and routine behaviors.

Integrating and Interpreting Health Data Through Professional Collaboration

As monitoring technologies expand, they provide continuous measurements that complement the periodic data obtained during clinical visits. Real-time tools make it possible to observe how glucose levels, heart-rate variability, sleep stages, body temperature, and other physiological signals change across daily routines. These measurements add insight into short-term and day-to-day fluctuations.

More data does not guarantee meaningful conclusions. Without clinical or situational context, continuous metrics can be overinterpreted or assigned undue weight. A brief rise in glucose, a night of disrupted sleep, or a low HRV reading may fall within normal physiological ranges and, by itself, indicate little about underlying health.

Clinical interpretation helps distinguish expected fluctuation from findings that warrant closer attention. Clinicians can compare real-time readings with reference ranges, consider the influence of medications or existing conditions, and assess whether changes recur in ways that align with reported symptoms. A measurement that appears unusual on first inspection may still reflect normal limits, while repeated deviations can signal the need for further evaluation.

In practice, real-time data becomes one component of a broader clinical assessment. When combined with history, symptoms, physical findings, and laboratory results, these measurements help clarify how the body responds to meals, activity, stress, sleep, temperature shifts, and other daily conditions. Continuous data can highlight recurring changes that merit follow-up and support more informed discussions between patients and clinicians.

AI-Driven Platforms and the Expanding Analysis of Health Data

Artificial intelligence is now used to process large health-related datasets, including information from wearables, biomarkers, imaging studies, and genetic testing. These systems apply statistical models to group similar signal profiles, flag values that fall outside expected ranges, and identify associations that may be difficult to detect through manual review. In this role, AI functions as a computational tool that organizes complex information and highlights results that warrant closer examination.

The reliability of AI output depends on the quality of the inputs on which these systems rely. Inaccurate sensor readings, incomplete records, or training material that does not reflect the populations being studied can produce false associations or obscure meaningful signals. Validation and periodic recalibration are therefore used to assess how well a model performs across settings and as new information is introduced. Privacy and security remain central concerns, particularly when biometric or genetic data are involved. Ongoing research also examines model bias, uneven performance across demographic groups, and limited transparency in how some algorithms generate their outputs.

As these tools develop, they are being applied in both research and personal data management. Some systems organize continuous streams of physiological information from wearables or home-based devices. Others support large-scale studies by analyzing extensive collections of anonymized health records. A more detailed discussion of AI's capabilities, limitations, and regulatory considerations appears in the chapter devoted to artificial intelligence later in this book.

Real-Time Stress Monitoring and Physiological Signals

Stress-related physiology shifts throughout the day in response to workload, temperature, noise, sleep quality, and other routine conditions. Wearable neurofeedback devices and real-time stress monitors record signals such as heart-rate variability, EEG activity, and markers of sympathetic activation. Together, these measurements reflect changes in autonomic activity, alertness, and recovery that can be reviewed alongside daily events.

In practical use, these readings are sometimes considered alongside practices such as paced breathing, meditation, or adjustments in schedule or workload. Such practices have been examined for their relationships with autonomic regulation and perceived stress, but their effects vary and depend on individual circumstances. Real-time monitoring adds context that can be interpreted in relation to personal history, daily conditions, and, when appropriate, clinical input. Rather than assigning meaning to moment-by-moment fluctuations, the emphasis is on understanding how stress-related physiology responds over time to routine demands.

Psychological and Behavioral Biohacking: Technologies for Exploring Cognitive and Emotional Dynamics

Biohacking now includes tools that measure cognitive and emotional activity in addition to physical metrics. These technologies record changes in attention, mood, stress-related signaling, and task performance under defined workloads, sensory conditions, or environmental settings. Their role is to capture data rather than interpret or classify experience.

Neurofeedback is one of the more established examples. EEG systems record electrical activity across frequency bands such as alpha, beta, and theta, which shift during tasks, rest, or exposure to stress. The resulting signals—numerical values or graphical traces—can be reviewed alongside observed behavior or self-reported experience.

Transcranial electrical stimulation (tES) and transcranial magnetic stimulation (TMS) offer additional ways to probe neural function. tES delivers low-level currents through scalp electrodes, while TMS uses magnetic pulses to induce activity in targeted cortical regions. In research settings, these techniques are used to examine how specific neural circuits respond to task demands or controlled stimulation, helping clarify the limits and variability of neural adaptation.

Artificial intelligence is used to analyze performance on adaptive cognitive tasks. These platforms adjust task difficulty or sequence based on user responses and record reaction time, accuracy, error rates, and measures of sustained performance. They organize large volumes of performance data without assigning causal explanations for observed variation.

Biofeedback presents physiological signals in real time. Devices measuring heart rate, heart-rate variability, muscle tension, breathing rate, or skin conductance show how autonomic activity shifts during tasks or periods of stress. Users may view these signals alongside techniques such as paced breathing or posture adjustment, though responses differ across individuals and require interpretation within context.

Virtual and augmented reality systems create controlled environments for studying navigation, attention, task engagement, or reactions to specific sensory or situational challenges. In research or clinical contexts, these systems are sometimes used under supervision to observe how individuals react to defined exposures.

Taken together, these tools allow cognitive and emotional activity to be measured under clearly specified conditions. When considered alongside behavior, symptoms, and environmental factors, they contribute to a more precise understanding of how mental processes adjust in response to different forms of input.

Personalized Nutrition and Gut Microbiome Analysis

Advances in sequencing technologies now allow more detailed measurement of the microbes that inhabit the gut. These organisms contribute to digestion, nutrient availability, and the production of short-chain fatty acids (SCFAs) such as butyrate, acetate, and propionate. Research has examined how differences among specific bacterial groups relate to glucose regulation, gastrointestinal symptoms, immune activity, and longer-term metabolic measures. These relationships remain an active area of investigation.

The microbiome interacts with multiple physiological systems. Microbial metabolites influence immune signaling, and communication between the gut and brain involves the vagus nerve, endocrine pathways, and cytokine activity. Studies examining microbial diversity and relative species abundance have linked these features to variations in inflammatory markers, bowel function, and self-reported mood. Taken together, this work suggests that microbial composition may help connect gut activity with inflammatory signaling, digestive function, and aspects of emotional state.

Microbiome sequencing is widely used in both research and consumer testing. Sequencing results identify which organisms are present in a sample and their relative proportions at a given time point, describing composition rather than functional activity. People sometimes compare results across time to observe how dietary changes, fiber intake, travel, illness, or medications correspond with shifts in microbial makeup. Such comparisons document change over time rather than define specific physiological effects.

The practical value of microbiome data depends on context. Dietary suggestions based on sequencing are best considered alongside symptoms, medical history, medications, lifestyle factors, and other laboratory findings. Healthcare professionals can help determine whether observed differences represent meaningful trends or fall within expected ranges. As research advances, the field continues to clarify how microbial composition relates to digestion, immune function, and metabolic processes.

Genetic and Epigenetic Testing: Perspectives on Inherited and Acquired Biology

Genetic testing identifies specific DNA variants that have been studied in relation to traits linked to aging, including cardiovascular function, lipid and glucose metabolism, and neurodegenerative disease research. Reports typically list single-nucleotide polymorphisms (SNPs) or other variants and indicate their frequency in population datasets. Examples include APOE ε4, examined in relation to Alzheimer's disease research, and MTHFR variants, which influence aspects of folate metabolism and methylation. These findings describe associations observed across groups. A variant may shift probability within a research cohort without determining whether or how a trait appears in an individual.

Epigenetic testing examines chemical modifications that influence gene activity without altering the DNA sequence. Most commercial tests measure DNA methylation at selected CpG sites using blood samples. Methylation values can vary with diet, physical activity, stress exposure, sleep quality, and environmental conditions. Some algorithms use these profiles to estimate cumulative biological change, often reported as an "epigenetic age." Results differ across laboratories and methodologies, and interpretation is constrained by tissue specificity and limited understanding of what changes at individual CpG sites represent.

Genetic and epigenetic information serve distinct roles. Genetic variants are inherited and remain stable across the lifespan, while epigenetic marks shift with time and may reflect recent or ongoing physiological states. Neither type of test specifies actions related to diet, supplementation, or medical care. Instead, they provide additional data points that can help frame understanding when reviewed alongside symptoms, medical history, medications, laboratory results, and lifestyle factors.

The usefulness of these findings depends on how they are interpreted. Clinicians and researchers consider whether a reported variant has established significance, whether an epigenetic estimate aligns with other indicators of health, and whether observed differences fall within expected ranges. Many results summarize population-level trends rather than individual trajectories. As research progresses, patterns of genetic variation and epigenetic change continue to refine understanding of how inherited structure and adaptive processes relate to aging.

A later chapter in this book reviews newer sequencing tools, computational approaches, and emerging applications that are shaping current work in genetics and epigenetics.

Environmental Biohacks: Interactions Between Biology and External Conditions

Environmental biohacking examines how specific external conditions—such as light exposure, temperature shifts, air-quality components, and contact with physical surfaces—relate to observable physiological changes. Some approaches are grounded in well-characterized physiological responses, while others remain exploratory and are investigated primarily under controlled research conditions.

Deuterium-depleted water (DDW) contains lower concentrations of the hydrogen isotope deuterium than standard water. Early investigations have explored whether reduced deuterium levels influence mitochondrial activity by measuring oxygen consumption and ATP production in cell and animal models. Human evidence remains limited, and available results do not demonstrate consistent or reproducible physiological effects. High cost and the lack of large, well-controlled studies confine DDW largely to experimental settings.

Circadian lighting systems adjust light wavelength and intensity to approximate natural day—night patterns. Research commonly examines melatonin release, alertness, sleep-onset timing, and shifts in circadian phase to understand how controlled lighting influences sleep—wake organization. These systems are used to investigate how patterned light exposure affects biological timing in contexts such as shift work or seasonal variation, rather than as standalone interventions.

Negative ion research focuses on charged airborne particles found at higher concentrations near waterfalls, coastlines, and certain natural environments. Investigations have measured air particle density alongside self-reported mood and physiological variables such as heart rate or breathing rate. Results remain inconsistent, and no established biological mechanism explains how negative ions would exert direct effects. The presence of multiple environmental factors in natural settings makes it difficult to separate any potential ion-related influence from broader contextual conditions.

Grounding, or earthing, involves direct skin contact with soil, sand, or other natural surfaces. Some small studies describe changes in heart-rate variability, skin conductance, or perceived calmness, but sample sizes are limited and methodologies vary widely. As a result, current evidence does not support reliable or reproducible physiological effects, and interpretations remain provisional.

Electromagnetic field (EMF) exposure from electronic devices is evaluated using regulatory benchmarks such as specific absorption rate (SAR) and frequency limits. Research has examined whether particular frequencies or intensities produce measurable biological responses, including changes in tissue heating or cellular signaling. Findings are mixed, and typical consumer exposures remain well below established safety thresholds. Interpretation depends on measurement methods and the broader exposure context.

Environmental biohacking does not replace foundational health practices such as physical activity, nutrition, and sleep. Instead, these approaches provide additional ways to observe how external conditions shape physiological responses. Their value lies in clarifying how environmental signals are registered by the body, contributing observational insight rather than acting as interventions.

Ethical and Privacy Considerations in Health Data Collection

As health-tracking technologies expand, they generate large volumes of physiological, biometric, and genetic material. Wearables produce continuous streams such as heart-rate variability, glucose readings, movement logs, and sleep estimates. Biomarker tests add periodic measures of hormones, lipids, inflammatory markers, and other laboratory values. Genetic analyses identify inherited variants that remain stable over time. Each category introduces distinct privacy concerns, especially when multiple sources are linked.

Ethical use depends on transparency about how records are collected, stored, shared, and protected. Many digital platforms now combine wearable outputs, laboratory results, app-based logs, and sometimes genomic profiles into unified accounts. When aggregated, these records can be re-identified even if individual fields are anonymized. Risks include unauthorized access to cloud-stored material, commercial resale to brokers or advertisers, and sharing with insurers or employers without clear disclosure.

Security practices help limit these risks by addressing how information moves, where it resides, and who can access it. Encryption protects content during transfer and storage so intercepted or improperly accessed files remain unreadable. Access controls restrict handling to authorized users, while de-identification methods such as hashing reduce the likelihood that records can be traced back to a specific person. Clear data policies also shape user choice, including how long records are retained, whether they can be deleted, and whether they are used to train analytic models. Regulatory frameworks such as HIPAA or GDPR address parts of this landscape, though many consumer health technologies operate partly or entirely outside their scope.

Analytic systems, including AI tools, process large collections of records to identify statistical associations or anomalies. Their performance depends on data quality, model design, and validation standards. Measures such as HRV, sleep staging, or glucose variability can be distorted by sensor error, movement artifacts, or incomplete capture. Genetic models may perform unevenly in populations underrepresented in training datasets. Results therefore require careful interpretation rather than stand-alone conclusions.

Responsible use of health tracking depends on consent, governance, and respect for the limits of analysis. Physiological measures, biomarkers, and genetic findings gain meaning only when considered alongside medical history, symptoms, medications, and environmental conditions.

Individuals benefit from understanding how their records circulate and from retaining meaningful control over their use. In this setting, privacy and security are not ancillary concerns but core requirements for any system that collects or analyzes personal health information.

A Vision for the Future

Biohacking is not a pursuit of extreme longevity claims or attempts to bypass the limits of biology. It is a structured form of self-experimentation that uses measurable inputs—such as light, temperature, movement, and sleep timing—to observe how the body responds and to adjust daily routines based on those observations. Its value lies in systematic inquiry: making controlled changes, examining variability, and distinguishing well-established findings from early-stage research.

Advances in measurement tools and computational analysis have expanded what can be assessed in everyday settings. These developments allow closer examination of how specific environmental and behavioral factors align with physiological processes involved in aging. Many of the practices discussed in this chapter remain preliminary.

Biohacking also reflects changes in how people interact with personal health data. People now track heart-rate variability, sleep, glucose trends, and other physiological signals, compare measurements over time, and use those comparisons to frame questions about sleep, activity, nutrition, stress, or environmental exposures. This engagement does not alter biological constraints, but it does create space to identify trends, detect inconsistencies, and approach health-related decisions more deliberately.

A consistent principle remains: observation guides inquiry while preserving the role of clinical judgment and established medical care. Biohacking tools can document how physiology changes under defined conditions without predicting outcomes or determining treatment. Used appropriately, these tools clarify how the body responds to particular forms of exposure.

The future of biohacking lies in disciplined measurement, clear boundaries, and careful interpretation. Its contribution is not in promising transformation but in offering practical ways to examine physiology as it changes over time. These practices help frame decisions while remaining grounded in what current evidence supports—and what it does not.

PART IV

Next-Generation Longevity— Personalized Health and Revolutionary Innovation

"The best way to predict the future is to create it."
—Peter Drucker

Aging research is entering a new era. Improvements in genomic sequencing, cellular imaging, and computational modeling now reveal biological changes associated with aging that earlier techniques could not detect. Progress in diagnostics, data analytics, and cellular investigation continues to expand what can be examined, bringing the workings of aging into sharper scientific focus than ever before.

Changes that once escaped detection—such as early shifts in inflammatory activity, mitochondrial function, or cellular maintenance—can now be identified, and modern biotechnology allows these processes to be examined with greater precision. Aging does not progress at the same rate throughout the body: neurons may alter their structure or signaling long before muscle fibers show measurable decline, and immune cells may shift toward more reactive states years before changes appear in metabolic tissues. Aging remains part of human life, yet current findings continue to show how unevenly these changes emerge, inviting closer examination of how later years unfold.

These advances also raise concrete ethical and practical issues. Technologies that track heart rate, sleep cycles, glucose trends, or other physiological signals are now widely used, bringing questions about data ownership, security, and access into direct view. The growing use of AI systems to process personal health information underscores the need for clear rules about how data are stored, shared, and reviewed. In this environment, progress depends not only on technical refinement but on transparent, accountable practices that protect the people whose information is being collected.

The chapters ahead examine developments that continue to change how aging is investigated, including genomic sequencing, high-resolution cellular imaging, and metabolic profiling. These practices make it possible to study aging at the level of DNA changes, alterations in cell structure, and shifts in energy use—details that earlier equipment could not capture.

Innovation cannot stand on its own. Its value depends on how people use these developments in daily life and how they work with clinicians to understand what the findings mean within care decisions and long-term planning. This leads into the closing chapters, which outline ways to form a Longevity Team and develop a personalized Longevity Plan built on informed conversation, shared decision-making, and ongoing attention to one's own biology.

The pursuit of longevity reaches beyond the hope for more time. It carries a wish to understand how the body changes and to act with greater attentiveness as those changes unfold. As scientific knowledge expands, we stand at a moment when aging can be examined with far more detail than any previous generation could access. The years ahead will reflect not only this expanding knowledge but the choices made in response to it.

CHAPTER 19

Regenerative Medicine: Emerging Pathways in Human Repair

"What lies behind us and what lies before us are tiny matters compared to what lies within us."
—RALPH WALDO EMERSON

Medicine is entering a decisive period of change. The long-standing model centered on symptom management is giving way to one that examines how biological systems change over time. Regenerative medicine stands at the center of this transition. Rather than relying on external treatments alone, it concentrates on the body's repair mechanisms—how cells respond to injury, how tissues rebuild, and how aging alters these processes. This shift draws attention to the conditions that support or limit recovery with advancing age.

The field is still emerging, marked by unresolved scientific questions, ethical concerns, fast-moving technological advances, and changing regulatory expectations. These forces create room for progress but also impose clear limits on what can be done today. In this chapter, we examine the foundations of regenerative research, outline current innovation, and consider the unanswered questions that continue to guide its direction.

IOULIA HOWARD AND DON HOWARD

The Science of Regeneration

Aging alters the body's capacity to repair itself. Rates of cellular division decline, tissues respond to stress less efficiently, and long-standing repair processes take on new characteristics. Three biological drivers help explain these shifts: cellular senescence, chronic inflammation, and changes in stem-cell activity. As discussed in Chapter 3, these elements form a central part of the biology relevant to regenerative medicine.

Senescent cells no longer divide but remain metabolically active, releasing cytokines and other compounds associated with inflammatory signaling. With advancing age, persistent low-grade inflammation can disrupt normal cellular communication and alter how tissues respond to injury or physiological stress. Stem cells, which contribute to tissue repair and turnover, may decline in number or show reduced responsiveness over time, further limiting the body's capacity to maintain or restore function.

Regenerative medicine draws on several active research areas, including stem-cell biology, gene-editing platforms, tissue engineering, and investigations of cellular senescence. Work in these domains examines how cells function at different stages of life, how genetic changes accumulate, and how tissues can be replicated or supported in laboratory conditions. These efforts expand understanding of biological processes that may guide future medical strategies.

Regenerative research is influenced by the same factors that affect everyday health, including preventive care, nutrition, and habits that alter inflammation, metabolic balance, and tissue repair. These physiological conditions set the limits within which regenerative findings can be applied in clinical settings, because they determine how well cells recover and how tissues function as the body ages. Much current work remains in laboratory and early clinical stages, with studies focused on describing age-related biological changes and identifying underlying mechanisms.

Stem Cell Therapies: Perspectives from Regenerative Biology

Stem cells occupy a central place in regenerative biology because they can generate multiple cell types during development and repair. The major categories include embryonic stem cells, adult stem cells, and induced pluripotent stem cells (iPSCs). Each enables investigators to examine how tissues form, remodel, and are maintained.

Embryonic stem cells can differentiate into nearly any cell type, but their use requires obtaining cells from early-stage embryos, which raises ethical concerns for many researchers and members of the public. Adult stem cells—present in bone marrow, adipose tissue, and other organs—support routine tissue maintenance and repair. Induced pluripotent stem cells demonstrate that mature cells can be reprogrammed into a pluripotent-like state, making it possible to study patient-specific cell behavior without relying on embryonic sources.

Stem-cell research spans several scientific domains. In orthopedics, studies examine how stem cells behave in cartilage, bone, and joint environments. In cardiovascular science, experiments measure their interactions within cardiac tissue models. In neuroscience, research groups map how stem cells differentiate into specific neural lineages and how these differentiation trajectories correspond with laboratory models of neurodegenerative change. Parallel work involves retinal, pancreatic, and other tissue systems, each revealing distinct cellular responses to aging and physiological stress.

Biological constraints limit current work in stem-cell research. Immune reactions, variation in cellular behavior, and the risk of uncontrolled proliferation all determine how differentiation protocols, biomaterials, and delivery methods are designed and tested. Techniques such as CRISPR and computational modeling can identify gene-level and cellular changes involved in development and repair, but these findings remain experimental.

Stem-cell research contributes to regenerative medicine by clarifying how cells divide, differentiate, and respond to physiological stress.

Studies in this field document age-related shifts in intercellular signaling, reduced repair capacity, and changes in tissue function. Taken together, these findings offer a more precise account of how aging affects cellular behavior.

Alongside direct cell-based approaches, attention has increasingly turned to the signaling mechanisms through which cells influence surrounding tissues.

Extracellular Vesicles and Exosome Therapy

Stem-cell research has contributed substantially to regenerative biology, but an adjacent area has expanded rapidly: the study of extracellular vesicles, including exosomes. Exosomes are nanometer-scale vesicles released by cells that contain proteins, RNA, and other molecular cargo involved in intercellular communication. They move through tissues via known transport pathways and participate in physiological signaling without introducing living cells.

Exosomes carry protein and RNA cargo that can influence intercellular communication, inflammatory activity, and the conditions under which tissues respond to stress or repair. Mesenchymal stem-cell—derived exosomes have been evaluated in models of tissue injury, neural cell systems, and skin biology to examine how vesicle-mediated signals affect local cellular behavior in these settings. Other studies assess whether exosomes can deliver defined molecular cargo under controlled conditions. In oncology, engineered vesicles are evaluated for their ability to deliver specific molecules to selected cell populations.

Exosome research is limited by challenges in production, characterization, and consistency. Generating vesicle preparations at scale is difficult, and current methods for identifying their contents and surface markers vary across laboratories. Stability, purity, and reproducibility differ among isolation techniques, making it hard to determine how specific vesicle populations behave in experimental systems. These uncertainties require tightly controlled methods and clear regulatory standards, since inconsistent preparations can produce conflicting or misleading biological results.

Exosomes transfer molecular cargo that modifies signaling within and between tissues. They appear in pathways associated with aging, injury response, and tissue maintenance, reflecting changes in vesicle-mediated communication under these conditions. These signaling changes are associated with measurable differences in cellular behavior as aging advances.

Regenerating the Human Body: Tissue Engineering and Bioprinting

Tissue engineering and bioprinting represent an expanding area of regenerative research, giving investigators ways to study how human tissues form, organize, and maintain function. These methods integrate biological insight with engineering techniques such as 3D bioprinting, bioactive scaffolds, and organoid systems, creating controlled environments in which the structural and functional properties of tissues can be examined in detail.

Creating tissues in the laboratory requires re-establishing elements of the body's architecture. Engineered tissue models use biocompatible scaffolds, living cells, and defined signaling cues to reproduce selected features of native structures. In skin research, for example, these models now include components such as pigmentation and glandular elements, allowing closer study of how tissues develop and respond to changing conditions.

Building tissues that function like whole organs remains a distant goal. Even with the spatial control provided by 3D bioprinting, reproducing organ-level complexity depends on understanding stem-cell behavior, gene-expression control, microenvironment design, and bioreactor conditions. Mechanical, metabolic, and electrical activities must align within the same tissue model, and current laboratory systems demonstrate how difficult it is to bring these activities together.

Organoid systems provide complementary ways to study organ biology. These small, three-dimensional cell assemblies reproduce selected aspects of organ development and function, giving researchers controlled settings to examine signaling pathways, tissue organization, and early disease processes.

Xenotransplantation uses gene-editing tools such as CRISPR to modify donor tissues and reduce immune reactions between species. Work in this area concentrates on altering surface markers, adjusting antigen profiles, and testing compatibility in experimental transplant settings. Progress remains preliminary, constrained by technical limits and long-established ethical questions.

Decellularization removes cells from donor organs and leaves the extracellular matrix intact as a structural guide for incoming cells. When recipient-derived cells are introduced into these matrices, their behavior reflects both the organization of the matrix and the intrinsic properties of the cells introduced.

Tissue engineering continues to face limits in scale, stability, and functional maturity. Forming stable vascular networks within larger engineered tissues remains particularly challenging, since cells rely on steady nutrient flow and efficient waste removal to survive. Efforts to meet these demands draw on angiogenic materials, microfluidic channels, and controlled bioreactor environments to sustain nutrient transport and maintain viable tissue. As these approaches become more complex and sensitive to manufacturing conditions, questions about production standards and equitable access arise alongside technical development, linking scientific progress with ethical and regulatory considerations.

Techniques such as bioprinting, scaffold design, organoid systems, and decellularized matrices give researchers controlled ways to trace how engineered tissues develop structure, endure mechanical stress, and sustain essential functions. When these systems are studied together, they show the conditions that support growth, hinder repair, or alter function as tissues mature. Taken together, this work offers a detailed view of how biological systems adapt through life, revealing connections between early developmental processes and the changes observed in aging tissues.

Bioelectric Medicine: Exploring Electrical Signaling in Regenerative Science

Electricity runs through nearly every biological system—from the regular pulse of the heart to the exchanges that allow neurons to communicate. It governs movement, influences cellular behavior, and contributes to the processes through which tissues assemble and maintain structure. Bioelectric medicine, an expanding field at the intersection of biology and engineering, studies these native electrical patterns to examine how tissues respond under controlled conditions.

Neural pathways have emerged as a central focus of this line of investigation. Controlled electrical stimulation allows investigators to observe how circuits respond to defined inputs, revealing aspects of the nervous system's capacity for reorganization. Work in spinal cord injury models and in conditions such as Parkinson's disease shows how electrical cues may correspond with shifts in neuroplasticity. Taken together, these findings indicate that neural systems can revise their activity in response to new demands.

Outside the nervous system, related research is examining how electrical fields intersect with broader cellular environments. Current investigations follow how these cues correspond with stem-cell activity, how wound settings respond in experimental models, and how skin cells migrate or assemble in laboratory systems. This expanding line of inquiry suggests that electricity participates in cellular organization across a wider range of tissues than previously appreciated.

As bioelectric medicine develops, it stands alongside other emerging disciplines—stem-cell research, tissue engineering, and gene-editing technologies—as part of a broader effort to understand how cells communicate and coordinate. The minimally invasive nature of many electrical techniques has drawn interest in how controlled electrical cues can be used to study cellular dynamics and track how living systems adjust activity in response to controlled electrical cues.

What once seemed like speculation—the idea that electrical signaling could deepen understanding of tissue behavior—has become a serious scientific pursuit. Although much remains unknown, the trajectory of this field reflects a growing commitment to mapping electrical signaling within the body and to examining how those signals contribute to the complexity of living systems.

Ethical and Practical Considerations

As regenerative medicine advances, it carries a range of ethical, practical, and philosophical questions. One of the earliest challenges is access. Many emerging interventions remain costly at first, raising the risk that only a small portion of society will be able to make use of them. These disparities point to concerns about fairness—who can learn about new developments, who can take part in early studies, and who may one day benefit from the knowledge gained. Ensuring that opportunity does not concentrate around a privileged few is central to how the field grows and how it serves the needs of diverse communities.

Ethical questions in regenerative medicine reach well beyond issues of cost. Work involving embryonic stem cells, gene modification, and other experimental methods raises concerns about study design, risk management, and how unintended effects should be addressed. Meeting these responsibilities requires regulatory systems that are both clear and consistently enforced—systems that support scientific progress while maintaining public confidence. Without such guardrails, it becomes difficult to judge when an experiment rests on sound reasoning and when restraint is warranted.

Public understanding also affects how regenerative medicine takes root. The pace of discovery often surpasses the pace of explanation, leaving room for confusion and unrealistic expectations. Clear accounts of both the science and its uncertainties help people interpret new findings with greater steadiness and reduce the likelihood of misunderstanding. In this sense, communication becomes part of the ethical responsibilities that accompany scientific work.

Looking ahead, the task is to hold scientific ambition alongside ethical care and fair access. Regenerative research may widen the questions that can be asked about living systems, yet its influence will depend on the choices made as the field develops. Decisions about study design, risk management, and participation will shape how this work is understood and who is able to benefit from it. The balance struck in these formative years will guide how future generations interpret this moment in biology and its implications for aging.

The Future of Regenerative Medicine— Pushing Scientific Frontiers

Regenerative medicine stands at a turning point. Work that once focused mainly on symptom management now reaches further, examining how living systems organize, adjust, and interact across time. This chapter has followed the sweep of emerging investigations—from precise stem-cell differentiation models and exosome-mediated communication to advanced methods in tissue engineering, bioprinting, and biomaterials. Further lines of research, including studies of bioelectric signaling, xenotransplantation, and gene editing, reveal a field moving at an accelerating pace and raising scientific questions that would have been difficult to imagine a generation ago.

Progress brings challenges that are both technical and ethical. Investigators continue to confront the complexities of cellular behavior, immune variability, and the difficulty of establishing stable blood vessels within engineered tissues. Another obstacle lies in determining how laboratory-built models correspond to the physiology of intact organisms—a gap that remains central to the field's long-term prospects. Ethical concerns move in parallel: the sourcing of cellular material, the limits placed on gene modification, and the question of who gains access all call for careful judgment and transparent explanation from those guiding the research forward.

Looking ahead, the central task is to balance scientific ambition with ethical care and fair access. Regenerative research is opening new ways to study how living systems change and adapt, yet its influence will depend on choices made as the field matures. Decisions taken in these years will affect how future generations understand this work and how it broadens our view of human biology and aging.

CHAPTER 20

Gene Therapy and Aging: Transformative Ideas in Genetic Science

"There is nothing in a caterpillar that tells you it's going to be a butterfly."
—R. BUCKMINSTER FULLER

Gene therapy has moved from scientific possibility to clinical reality, and its reach now extends into the study of how aging unfolds at the level of DNA. By examining how genetic sequences are maintained, repaired, or altered, this field sheds light on the microscopic events that occur within cells over a lifetime. Gene-based tools make it possible to follow how mutations arise, how repair systems respond to stress, and how changes in gene activity accompany advancing age. This chapter introduces these tools and considers their use in investigating aging at its genomic origins.

CRISPR and the Changing Horizon of Genetic Research

Few developments have altered genetic science as profoundly as CRISPR. First identified as part of a bacterial defense system, it has since been adapted into a tool capable of cutting or editing DNA with remarkable precision. This technology makes it possible to study gene function directly, build models of inherited disorders, and investigate genetic variations linked to aging biology at a level of detail that was previously out of reach.

CRISPR is also beginning to appear in controlled clinical settings. In neurology, it is used primarily as a research tool to modify variants in genes such as APP and PSEN1, allowing investigators to observe how specific mutations influence amyloid production and aggregation. This work clarifies which genetic changes contribute to disease biology and how they act, improving understanding of Alzheimer's disease mechanisms rather than serving as a diagnostic tool. In ophthalmology, CRISPR is delivered directly to the retina in early clinical trials for certain inherited disorders, allowing gene activity to be altered within the affected tissue. In cardiovascular medicine, clinicians are testing CRISPR-based editing of PCSK9 to alter long-term LDL regulation at its genetic source. Together, these examples show how gene editing is entering medical practice in carefully defined, tissue-specific, and measurable ways.

Newer gene-editing methods, such as base editing and prime editing, change a single DNA letter at a time instead of cutting the entire strand. Working at this smaller scale reduces the chance of unintended changes and allows a specific nucleotide to be adjusted without disturbing nearby sequences. Computer models can now predict where a planned edit is most likely to occur, improving the ability to track off-target effects. These methods are still applied cautiously, but they signal a shift toward gene-editing techniques that are more precisely guided and easier to verify.

CRISPR's precision does not remove its risks. Edits can introduce unintended genetic changes, and the long-term effects of modifying DNA in living tissues remain uncertain. Editing eggs, sperm, or early embryos raises additional concerns because any change made at that stage can be inherited. These issues underscore the need for strict oversight and open public discussion as gene editing advances. Within these constraints, CRISPR is used to explore how specific genes influence cellular pathways linked to aging.

Reprogramming Aging: Broader Gene Therapy Approaches

While CRISPR often receives most of the attention, aging research also examines how gene activity changes over time through epigenetic reprogramming. This work focuses on chemical tags on DNA that influence when genes are active or silent. A central tool in this field is a group of proteins known as the Yamanaka factors—OCT4, SOX2, KLF4, and c-MYC—which were first shown to shift mature cells toward an earlier developmental state. In aging research, these proteins are applied in limited amounts to observe how gene-expression patterns respond. They do not alter DNA sequences; instead, they modify regulatory signals that help define a cell's identity. Because these signals can affect many genes at once, their level and duration must be carefully controlled to preserve normal cellular function.

Synthetic biology builds genetic circuits. These circuits allow investigators to examine how cells regulate metabolism, organize into tissues, and respond to stress. In parallel, studies of immune signaling track how aging alters the messages immune cells send and receive, including cues released by senescent cells. Together, these approaches show how cellular communication changes with age and how different cell types interact.

Gene-directed strategies still face practical hurdles, including how to deliver genetic material to specific tissues, how to monitor long-term effects, and how to make these techniques usable outside specialized laboratories. Despite these limits, laboratory studies now allow more precise examination of how particular genes influence cellular function. When combined with CRISPR, this work shows that aging reflects changes occurring across many tissues rather than a single process. As these capabilities expand, they help clarify which genetic and cellular changes most influence the aging process.

Beyond Repair: Exploring Longevity-Associated Genes

Aging research is moving beyond the study of genetic damage to examine how certain genes support routine cellular function. Instead of focusing only on mutations that disrupt normal activity, current studies look at genes involved in repair, energy use, and responses to stress. Learning how these genes operate gives a clearer picture of how the body maintains itself and how those maintenance processes shift with age.

Several genes stand out because of the central roles they play in cell function. FOXO3 and SIRT6 help cells manage stress, repair DNA, and regulate energy use. Studies of the TERT gene track how it directs the production of telomerase, the enzyme that maintains chromosome ends. Reviewing these genes together shows how their activity changes with age and which conditions appear to influence those changes.

Parallel research examines genes involved in oxidative and inflammatory responses. NRF2 plays a major role in antioxidant defense and is assessed for how it affects a cell's response to oxidative stress. NF-κB is central to inflammatory signaling and is analyzed for its role in the persistent, low-level inflammation often seen with age. Studying these regulators helps clarify how oxidative stress and inflammation contribute to age-related cellular changes.

Focusing on these genes shifts aging research toward the internal processes that keep cells functioning and reveals how those processes weaken over time. This work identifies genetic changes closely tied to cellular aging and highlights areas where further investigation may be most informative.

Navigating the Challenges: Scientific, Ethical, and Societal Considerations

Gene therapy alters DNA inside living cells, a process that demands precise methods and careful safety checks. As these techniques advance, the procedures for introducing genetic changes, measuring their effects, and addressing unexpected outcomes must be clearly established and consistently applied. Progress in the field also increases the need to explain these methods in clear, accurate language to patients, clinicians, and the public, and to understand how genetic modification is viewed across different medical and social settings.

Maintaining accuracy in gene editing remains difficult. CRISPR can cut or modify DNA at chosen locations, but it can also create unintended changes, and the long-term effects of those edits are still uncertain. Base editing and prime editing reduce some of these risks by altering single DNA letters instead of breaking both strands, yet no technique can direct every outcome with complete reliability. These limits are especially relevant when gene-editing tools are used to study aging, where unplanned edits can make it harder to determine which changes are caused by aging and which arise from the editing process.

Ethical questions become especially complex when gene editing involves changes that can be inherited. Germline editing affects eggs, sperm, or early embryos, so any alteration may pass to future generations. This possibility raises practical decisions about who should guide such choices, how risks are evaluated, and how potential benefits are weighed. It also sharpens the distinction between correcting serious genetic disorders and modifying traits unrelated to disease.

Gene-therapy also raises practical questions about how people and communities respond to developments in genetic science. These changes may influence how families think about aging, arrange caregiving, and divide responsibilities across generations. As scientists learn more about physical and cognitive changes linked to aging, care systems may adjust how support is organized and how future needs are anticipated. Because these choices affect the delivery of care and the use of shared resources, guidance from medical and public agencies helps people understand options and plan accordingly.

Access to gene therapy remains uneven. If the use of these techniques is limited to people with significant financial resources, existing health and social disparities may widen. Ensuring fair access requires clear rules for offering gene-based interventions, monitoring their safety, and involving the public in decisions about their use.

The direction of gene therapy will rest on scientific progress and on the rules that govern its use. Policies that set boundaries, define safety requirements, and explain these standards to the public will influence how genetic modification is understood. Choices about where gene editing is permitted and where it is restricted will show how society balances hope, caution, and responsibility.

Gene Therapy and Aging: What's Coming Next?

Gene-therapy is entering a period of rapid development, supported by tools that allow closer study of how genes work together inside cells. Multiplexed gene editing reflects this shift. Instead of changing one gene at a time, multiplexed systems can adjust several genes in a single experiment, making it possible to see how a change in one process affects others. The MAGESTIC platform is one example: it produces multiple edits at once and records each change, enabling detailed study of DNA repair, metabolic regulation, and cellular stress responses in experimental models that connect cell-level changes to aging processes.

Combining gene-editing techniques with stem-cell studies allows direct examination of how specific genetic changes influence the growth and repair of tissues. New procedures for assessing stem-cell quality before use in experiments make it easier to track how genetic and epigenetic signals guide cell behavior as tissues form or are replaced. Used together, these approaches show how early cellular changes contribute to age-related shifts in tissues and organs throughout the body.

Artificial intelligence is increasingly used to manage the large datasets involved in gene-editing studies. Machine-learning programs can sort genomic and protein information to identify likely editing targets, flag positions where unwanted changes may occur, and support experiment design. When applied to aging research, these tools help reveal how a genetic change in a single cell can affect processes linked with structural integrity, repair, and resilience in the body's tissues.

Even with rapid progress, several practical issues remain. Long-term safety must be monitored, rules for ethical use must be clearly defined, and regulators must decide when gene-editing approaches are ready for early-stage clinical evaluation. Moving from laboratory studies to initial clinical settings also requires methods that can be carried out reliably and reproducibly. As these steps develop, gene-therapy is refining how aging is examined, by identifying which genetic changes influence aging, how those changes vary from person to person, and how they contribute to differences in the rate at which tissues weaken with time.

Final Thoughts: Gene Therapy and the Future of Aging Research

Gene therapy offers a clearer view of aging by making it possible to study how genes behave inside living cells. Earlier approaches often relied on outward signs—such as weaker tissues or slower repair—without access to the genetic activity that initiated those changes. With current editing and monitoring methods, scientists can now see when key signals shift inside cells, how those shifts alter cell function, and how early molecular changes may later manifest across tissues and organ systems.

Advances in genetic science are revealing how specific cellular activities change with age. Work in cellular reprogramming, targeted gene editing, and improved tissue models shows when repair slows, when stress-response systems weaken, and when energy use becomes less efficient. These findings help explain why some tissues tire or mend more slowly than others and offer clearer starting points for future investigation of conditions that become more common with age.

New genetic techniques also make it possible to study inherited mutations and tissue-specific vulnerabilities that were once difficult to analyze. By pinpointing changes in DNA that affect repair, energy use, or stress responses, these methods help explain why certain tissues show early strain or respond differently to injury. They also reveal molecular shifts that appear long before outward symptoms, giving science a clearer map of how aging progresses across diverse tissue systems.

Work at the genetic level raises practical and ethical questions that cannot be ignored. Editing DNA in adult tissues requires defined limits on risk and continued monitoring. Editing genes in eggs, sperm, or early embryos carries added weight because any change can be inherited. These issues raise concrete decisions about who should have access to genetic techniques, how safety is judged over time, and how to distinguish between correcting harmful conditions and altering traits that do not impair health.

Even with these challenges, gene-therapy is clarifying where aging begins at the cellular level. Examining genetic activity this closely raises questions central to aging biology: which early shifts in gene regulation mark the onset of cellular decline, and why do some tissues exhibit these shifts long before others? The answers do not promise escape from time; they define the boundaries within which living matter changes, repairs, and deteriorates. In this way, the study of genetic aging offers a more exact account of how time acts at the cellular and tissue level.

CHAPTER 21

Technology and Longevity: Tools Changing How We Age

"Any sufficiently advanced technology is indistinguishable from magic."
—ARTHUR C. CLARKE

New technologies are widening the ways aging and healthspan can be observed, measured, and examined. Artificial intelligence can sift through immense physiological datasets; CRISPR and related methods allow targeted changes at the level of DNA; nanomaterials operate at scales small enough to interact with cells; and robotics brings steady, exact motion to surgical and rehabilitative settings. Tools that once lived only in experimental laboratories now help document how tissues recover, how cells respond to stress, and how physical function changes over the course of aging.

Other innovations extend these capabilities into daily life. Wearable devices track heart rhythm, glucose variation, sleep timing, temperature, and movement throughout the day; digital systems create physiological "twins" that run simulations based on real data; bioelectronic sensors register electrical signals within nerves and tissues; and virtual or augmented environments offer controlled spaces for rehabilitation, exercise, and behavioral study. As these technologies continue to expand, this chapter considers how they may influence decisions about healthspan—how information is collected, how strategies are planned, and how everyday routines may adapt as new tools enter common use.

Wearable Technology: A New Era of Physiological Observation

Wearable technology has progressed far beyond simple step counters, becoming a set of tools that record continuous physiological measurements throughout the day. Smartwatches, fitness bands, and biometric rings now track heart rate, sleep timing, activity levels, skin temperature, and indicators of stress such as heart-rate variability. These devices provide a longitudinal record of how the body functions during work, rest, and movement—information that intermittent clinical measurements cannot capture.

A notable step in this progression is the addition of ECG recording functions to consumer devices. With a brief touch, users can generate tracings of the heart's electrical activity for later clinical review, allowing irregular rhythms to be documented at the time they occur. Other sensors broaden the range of measurable features: optical and bioimpedance systems estimate blood pressure without a cuff, hydration monitors read bioelectrical or sweat-based signals, and continuous glucose monitors track glucose fluctuations throughout the day. Non-invasive glucose-monitoring methods are also under active development, while smart fabrics with embedded sensors and early implantable devices add new ways to follow temperature changes, muscle activity, posture, and movement as physiological conditions evolve.

Artificial intelligence analyzes the large volumes of data generated by wearables, sorting information on heart rhythm, sleep, movement, temperature, and glucose variation. These systems detect changes that unfold gradually over weeks or months—shifts that may be difficult to recognize without continuous measurement. As these capabilities expand, efforts to link wearable datasets with electronic health records have gained momentum, placing day-to-day physiological information alongside data gathered in clinical settings.

Viewed together, wearable devices offer a far more detailed account of physiology than measurements taken only during clinic visits. Instead of isolated readings, they trace how key signals rise and fall across the day. The value lies in the continuity itself: long stretches of data make gradual shifts easier to identify and place moment-to-moment variations in context. They offer a practical way to observe how the body responds to daily routines and to detect slow changes in basic physiological signals that occur without immediate physical cues.

CRISPR and Bioprinting in the Future of Aging Research

CRISPR has become one of the most precise tools for examining how genes influence aging. First identified as part of a bacterial defense system, it now enables targeted changes to specific DNA sequences, allowing scientists to test the role of individual genes and observe how those alterations affect cell behavior under defined experimental conditions.

Bioprinting and tissue engineering offer a practical way to study age-related processes at the level of cells and tissues. By layering biomaterials with living cells, these methods create small tissue replicas that capture selected features of real organs. These replicas let investigators explore how cells arrange themselves, how structures handle physical stress, how nutrients move through tissue, and how these features change with age. They provide a view of biological changes that cannot be directly observed in humans.

Together, CRISPR, tissue engineering, and bioprinting offer several vantage points on the biology of aging. Each reveals a different aspect of change—gene activity, cellular arrangement, or tissue response—providing a breadth of insight that would not be accessible through a single approach.

Nanotechnology in Medicine

Nanotechnology allows work to be carried out at dimensions where individual molecules and small molecular assemblies can be examined directly.

At this scale, scientists can follow chemical reactions inside cells, track how signals move between molecules, and study how the local environment of a cell affects its behavior. These methods now permit highly sensitive detection of molecular changes and provide clearer access to processes that once unfolded beyond the reach of laboratory instruments.

Some of the more speculative proposals in nanotechnology involve nanoscale devices—so-called "nanobots"—designed to move through the body and interact with cells directly. These ideas remain theoretical, but they help scientists consider what might be possible as fabrication and sensing methods improve. More immediate are nanosensors, which are engineered to detect biochemical signals at exceptionally low concentrations. These sensors are being studied for their ability to register early molecular changes that precede more visible physiological shifts, including those associated with aging.

Nanotechnology remains a young field, valued more for the access it provides to molecular activity inside cells than for any immediate clinical use. As measurement and fabrication methods advance, these tools make it possible to observe reactions, structural changes, and signaling events that were once too small to detect. At present, this work is exploratory, aimed at understanding how changes at this scale influence cellular behavior and the tissues cells form.

Digital Twin Technology: Emerging Models for Studying Biological Complexity

Digital twin technology creates a computer-based model of a person or a specific organ that updates as new data arrive. Instead of storing single readings in a chart, a digital twin integrates repeated measurements from wearables, imaging studies, laboratory tests, genomic analysis, and environmental records to keep the model aligned with the body it represents. As this record builds, the twin shows how markers such as heart rhythm, sleep timing, blood chemistry, or body composition change across days, weeks, and months—revealing recurring relationships among variables that are difficult to detect from occasional measurements alone.

Artificial intelligence plays a central role in building and updating digital twins. By processing large sets of physiological, metabolic, and structural data, AI systems can generate simulations that reflect how organs function, how energy is used, and how cells respond under different conditions. These simulations give scientists a practical way to explore questions that cannot be examined directly, allowing them to test ideas, compare possible outcomes, and study biological interactions before laboratory or clinical experiments.

Digital twin models are beginning to contribute to aging research as well. They allow scientists to examine how changes in diet, cellular senescence pathways, or epigenetic markers may correspond to shifts in metabolism, immune activity, or other measurable features of aging. These simulations help generate testable ideas and highlight relationships that are difficult to examine directly. In drug development, they offer a way to model how genetic differences may influence responses to experimental compounds before laboratory or clinical testing begins.

For many people, a digital twin offers a way to follow their own physiology over longer periods. Instead of relying on population averages, a twin draws on personal data to create simulations that reflect a person's sleep cycles, daily movement, heart rhythm, or metabolic markers. They illustrate how everyday habits correspond to measured biological changes across extended intervals.

Significant challenges remain—accuracy, privacy, security, and the ability to interpret results in a reliable way will all influence how digital twins develop. Even so, progress continues at a steady pace. It is not difficult to imagine a future in which a person has a virtual counterpart that updates alongside their own physiology, offering a way to follow long-term variations in metabolism, activity, or organ function. As this science advances, it may offer clearer insight into how metabolic, immune, and cellular processes change with age and how those changes differ among people.

Robotics: Technological Influence on Surgery and Rehabilitation Research

Robotic systems have moved from early prototypes to a regular presence in medicine, offering new ways to examine precision and movement in both surgery and rehabilitation. What once felt speculative now provides clinicians and investigators with stable platforms for studying how a surgeon's technique is carried out—the position of an instrument, the force applied, and the way each movement is translated through the robotic system.

In surgery, platforms such as the da Vinci system are now routine tools that make delicate, minimally invasive procedures easier to observe and analyze. Their articulated instruments and magnified, high-definition views show how hand movements are steadied, how tremor is reduced, and how small adjustments translate into precise action at the operative site. Comparisons with traditional techniques continue, with results that vary depending on the type of operation and the demands placed on the surgeon.

Robotic systems also play a growing role in rehabilitation research. Wearable exoskeletons from companies such as ReWalk Robotics and Ekso Bionics allow investigators to watch how a person shifts weight, initiates steps, and maintains balance as the device supports movement. In clinical laboratories, robotic gait-training systems such as the Lokomat create steady, repeatable conditions for guided walking, making it possible to follow how motion changes with repeated practice. These tools add structure and consistent measurement to the study of movement and recovery.

Taken together, these robotic platforms offer clearer ways to observe both surgical action and rehabilitation. As they continue to advance—and begin to operate alongside adaptive, learning-based systems—they may allow investigators to model how specific motor patterns are initiated, refined, and repeated, providing more detailed insight into the mechanics of human movement.

Beyond Mechanics: Bioelectronics and the Expanding Study of Physiological Signaling

Bioelectronics shifts attention from mechanical motion to the electrical signals that move through the nervous system. By interacting directly with neural circuits and their measurable activity, these tools offer a way to observe how nerves communicate and how electrical impulses travel within and between organs.

Vagus nerve stimulation, first developed for certain neurological and psychiatric conditions, has drawn attention for how electrical signals interact with inflammatory and stress-related pathways. Long-standing medical devices such as pacemakers, defibrillators, and deep brain stimulators show how electrical modulation already plays a role in clinical care, illustrating the range of responses the body exhibits under controlled electrical input.

Peripheral nerve stimulation directs electrical signals toward motor, sensory, or autonomic fibers when nerve pathways have been disrupted. Instead of restoring movement through mechanical assistance alone, this method works by activating neural circuits directly, revealing how they adjust their signaling when given targeted electrical input. These changes highlight the nervous system's capacity to reorganize when established patterns of communication are altered.

Although still developing, bioelectronics brings engineering and neuroscience into direct contact with the body's electrical activity. By offering tools that can record or deliver signals with increasing precision, it allows observers to see how electrical cues influence cellular and physiological processes. As these capabilities expand, they offer clearer views of how the body coordinates repair, movement, and function through integrated physiological pathways.

Expanding the Frontiers of Longevity Technology

When robotics, bioelectronics, and artificial intelligence converge, they form a growing set of tools that alter how aging and physiology are examined. AI-guided bioelectronic systems can adjust stimulation settings over time and track how neural circuits react under controlled conditions. Robotic rehabilitation platforms now incorporate adaptive feedback, creating stable settings in which balance, coordination, and gait can be observed with unusual detail. Wearable bioelectronic devices are evolving as well, recording measures such as heart-rate variability, temperature shifts, and features of neural signaling that may reflect age-related physiological change.

Robotic systems are also taking on a larger role in work involving older adults. They are being used to support mobility tasks and to help people take part in structured physical activity with steadier guidance and measurement. Augmented and virtual reality are moving into laboratory and clinical settings as well, where they assist with medical training, rehabilitation exercises, and studies of cognitive engagement. Together, these technologies create settings where physical movement, attention, and task performance can be examined with greater resolution.

Robotics, bioelectronics, and artificial intelligence mark a shift in how aging and physiology are investigated. Their value lies in the detail with which they expose processes that were once difficult to follow—how movements unfold, how signals are exchanged, and how tissues respond over time. As these tools advance, they offer additional ways to study aging in a technologically rich environment, one in which biological activity can be tracked more continuously and understood through both mechanical and computational perspectives.

Virtual and Augmented Reality: Immersion, Perception, and Behavior

Virtual reality (VR) and augmented reality (AR) create spaces where digital elements blend with or transform the surroundings a person sees. VR draws someone into fully crafted scenes, while AR places digital cues within the world already in front of them. Together, they offer new ways to explore perception, engagement, and the experience of moving through environments shaped by both physical and digital features.

VR's ability to create vivid, absorbing scenes has drawn particular interest. Many people describe its tranquil environments—shorelines, forests, underwater vistas—as offering a brief shift in attention. These impressions vary, highlighting VR's usefulness for observing how immersive settings influence attention, mood, and sensory experience.

VR can construct scenarios that place a person in demanding or unfamiliar situations, offering a way to watch how posture, movement, attention, or decision-making change as conditions shift. Because these environments can be modified in real time—made simpler or more complex, quieter or more stimulating—they allow observers to follow how someone adapts as one task gives way to another. Cognitive VR activities such as sequenced memory tasks, spatial-navigation exercises, and attention-focused challenges add further opportunities to see how different forms of engagement relate to cognitive performance under everyday conditions.

AR takes a different approach. By placing digital cues within real-world surroundings, it creates environments where memory tasks, spatial navigation, or step-by-step activities can be practiced with added guidance. In movement-based settings, AR overlays make posture, alignment, and motion easier to see as they unfold, supporting consistent conditions for observing how movement changes with repetition.

VR extends this approach by allowing everyday actions—reaching for objects, moving around obstacles, coordinating steps—to be practiced in interactive spaces that can be adjusted as needed. When visual or auditory feedback is added, these environments can hold a person's attention for longer periods, making it easier to follow how movement evolves with continued practice.

Together, VR and AR create environments where behavior, attention, and movement can be observed as they naturally unfold. They show how a person approaches a task, adapts to shifting cues, and interacts with surroundings that respond to each action in real time. By blending digital elements with the physical world—or replacing it altogether—they open a wider view of human experience, revealing how engagement deepens when the environment changes in response to the person moving through it.

Immersive Technology in Exercise Environments

Virtual and augmented reality are changing how people approach physical activity by creating immersive settings that alter the experience of movement. VR platforms draw participants into vividly rendered worlds—boxing rings, mountain paths, open oceans—where exercise blends with exploration and play. Programs such as Supernatural and FitXR use rhythm, choreography, and responsive visual cues to hold attention, turning routine workouts into experiences that feel more like games than obligations.

AR takes a different approach, placing digital cues within the real world. During yoga, Pilates, or strength-focused sessions, it can display prompts showing posture, alignment, or the flow of a movement sequence, offering visual guidance that helps people stay oriented within the routine. Many are now exploring how AR might support balance and mobility exercises, particularly for older adults, where clear visual cues can support steadier execution of movements during physical tasks.

Advancing Medical Training and Surgical Visualization

VR and AR are changing how medical education and surgical training are approached. Virtual environments let clinicians and trainees rehearse procedures repeatedly, studying anatomical structures and clinical scenarios with a level of control and safety that live settings cannot always provide. These simulations build familiarity with complex tasks and support a steady, deliberate approach to developing technical skill.

Augmented reality extends this capacity by placing digital information directly within the operative field. In most systems, this information appears as overlays—labels, outlines, or highlighted structures that sit on top of the real-world view—helping learners orient themselves within complex anatomical spaces. These elements offer real-time visibility of features that may be difficult to see unaided. By clarifying spatial relationships during preparation or imaging review, AR provides an added layer of visual context in medical education.

Together, VR and AR widen the spaces in which medical learning can occur. They offer settings in which difficult skills can be explored safely and repeated as often as needed, giving clinicians and trainees another way to gain familiarity with procedures while still relying on traditional instruction. In this sense, immersive technologies add depth to the preparation for medical practice, enriching how people train for the demands of clinical work.

At the Threshold of Technology and Aging

The future of longevity is emerging from the convergence of technologies that only recently felt out of reach. Advances in bioelectronics, nanotechnology, gene editing, immersive digital systems, and artificial intelligence are progressing side by side, each expanding the kinds of questions that can be asked about aging. Together, these tools offer closer views of biological change and give a clearer sense of how the body evolves gradually, often beneath everyday awareness.

Health monitoring is poised to move far beyond today's wearable devices. Physiological data may eventually flow into analytic platforms, digital-twin models, or individualized simulations that trace how the body changes over months or years rather than through occasional readings. Regenerative science is advancing alongside these technologies: bioprinted tissues, stem-cell methods, and cellular reprogramming efforts continue to reveal how aging cells adjust, persist, or decline. Taken together, these developments reflect a growing effort to examine aging as a collection of intertwined processes rather than as something governed by a single explanation.

Such developments also bring complexity. As societies adjust to shifting demographics and the growing presence of new technologies, questions around gene editing, intelligent systems, and the boundaries between humans and machines require careful attention. The future of longevity work will depend not only on what science makes possible, but on how thoughtfully and equitably these emerging capacities are brought into everyday life.

As these fields advance, they change the kinds of questions that can be asked about aging. Instead of centering only on decline, emerging technologies draw attention to autonomy, adaptation, and the everyday experience of later life. In this evolving era, longevity becomes less a distant aim and more a domain shaped by deliberate choices, ongoing inquiry, and the growing exchange between people and the technologies they bring into their lives.

CHAPTER 22

Artificial Intelligence and Aging: Promise, Precision, and Limits

"The only way to discover the limits of the possible is to go beyond them into the impossible."
—ARTHUR C. CLARKE

Historically, medicine often approached aging-related conditions reactively, intervening only after symptoms intensified and complications had already taken hold, while early signals often remained obscure or unrecognized. Today, artificial intelligence (AI)—a rapidly evolving field that enables machines to perform tasks associated with human cognition—is contributing to a shift in how such signals are studied.

AI is distinctive in how it processes information. At its core, AI encompasses techniques such as machine learning, in which algorithms examine large datasets to detect relationships and recurring features that traditional methods may not readily capture. Within machine learning, deep learning stands out for its use of neural networks—computational architectures loosely inspired by the human brain—to extract information arranged across multiple layers of data.

Through these capabilities, AI expands the analytical tools available to researchers and clinicians. It allows examination of physiological information with finer resolution, supports the construction of more detailed models, and enables study of how aging-related data change as biological processes shift. AI functions as a means of working with data at scales and dimensions beyond human inspection, offering a broader view of how age-related variation emerges and accumulates.

AI expands what can be analyzed and distinguished within biological and clinical data. Genomic information, metabolic measurements, imaging results, wearable-sensor outputs, and clinical records can be examined within unified systems capable of detecting relationships and features that exceed the reach of conventional methods. In imaging, AI-supported systems can draw attention to elements that the human eye may struggle to distinguish, strengthening interpretation while keeping clinical judgment at the center. Histopathology is undergoing a similar shift, as AI systems help standardize how microscopic tissues are reviewed, offering more consistent ways to assess the details present within each sample.

Research is advancing in areas such as AI-analyzed blood-based biomarker panels, where investigators study whether molecular traces in blood or other fluids can reflect biological change as time progresses. In genetics, polygenic risk scoring (PRS) uses computational models to evaluate how combinations of variants relate to inherited traits. These methods are most informative when interpreted alongside clinical evaluation and established scientific approaches, each adding context to the other.

Digital twin technology broadens the investigative tools available to researchers studying aging. By constructing virtual models of physiological systems, researchers can simulate changes in nutrition, physical activity, or medication exposure and observe how modeled responses develop. These simulations provide a controlled setting for testing hypotheses and examining how variables interact under defined conditions.

At the population level, AI integrates information from hospitals, wearable devices, and environmental monitors. During the COVID-19 pandemic, these platforms were used to examine transmission behavior, viral evolution, and demand on healthcare resources. Their use revealed both the reach and the limits of AI, underscoring that concerns about bias, privacy, equitable access, and interpretability remain central to responsible deployment.

By processing information at scales and combinations beyond human inspection, AI helps researchers see how physiological changes accumulate across the years. Its contributions gain strength when paired with established scientific methods and clinical judgment, each clarifying the other. Together, they offer a more complete view of aging biology than any single approach can provide.

From Molecule to Medicine: How AI Is Transforming Drug-Discovery

The search for new pharmaceuticals has always been marked by uncertainty. Many promising compounds falter as their biochemical behavior becomes clearer, and the intricacies of aging biology only increase this complexity. Artificial intelligence is beginning to reshape how researchers navigate this terrain. By scanning immense chemical libraries, modeling interactions at molecular scales, and identifying candidates worthy of closer examination, AI introduces new efficiencies while experimental validation remains essential.

AI's influence extends to drug repurposing as well. Medications such as metformin, rapamycin, or compounds studied in senescence research—first examined in fields such as endocrinology, oncology, and immunology—are now being revisited in aging biology. The strength of AI lies in its ability to sift through vast biomedical datasets and uncover associations that may prompt new hypotheses, helping researchers frame questions that guide subsequent laboratory work.

Clinical research is also beginning to incorporate AI-based tools. Machine-learning models can analyze clinical histories, biomarker patterns, or genetic information, offering ways to examine participant variability and refine trial design. Computational methods are being explored to study how adverse-event signals emerge or how early dose-range exploration might be structured. These developments underscore the need for rigorous oversight, equitable access, and careful integration as AI becomes more embedded in research.

AI is increasingly used in synthetic biology to design molecules and other investigative tools. It assists in creating novel molecules and engineered molecular systems intended to probe cellular pathways related to mitochondrial behavior, senescence, or autophagy. These efforts remain exploratory, but they reflect the expanding ambition of scientists seeking to understand aging through increasingly sophisticated methods.

AI and the Study of Biomarkers in Aging

Aging reflects shifts in molecular activity, alterations in cellular function, and changes in tissue physiology that accumulate over time. Chronological age marks the passing of years, while biological age captures measurable features—methylation signatures, protein profiles, metabolic signals—that indicate how the body is functioning beneath the surface. Because these processes operate at multiple biological levels, researchers increasingly turn to artificial intelligence (AI) to analyze genomic, proteomic, metabolic, imaging, microbiome, and wearable-derived data within a single analytical system. AI helps investigators examine biomarkers that reveal how physiological processes accumulate and interact.

Epigenetic clocks estimate biological age by modeling shifts in DNA methylation across many genomic sites, offering a molecular record of how the body changes. These clocks read methylation marks in a manner analogous to tracing landmarks, revealing how gene-regulatory signals adjust as cells respond to the passage of years. AI techniques refine these models by identifying methylation sites that carry the strongest interpretive weight, linking measured signals more closely to underlying biological states. Platforms such as Deep Longevity, Gero.ai, and Horvath's Epigenetic Clock use machine-learning methods to study how methylation signatures align with aging-related change, opening room for deeper examination of forces that steer cellular behavior.

Metabolic and proteomic analyses broaden the view of aging by tracing how molecules rise, fall, and regroup as the body responds to the demands of living. AI helps investigators discern recurring molecular features within these datasets, creating a clearer sense of how biochemical pathways adjust under shifting physiological pressure. Signals gathered from wearables—glucose fluctuations, changes in heart rhythm, the structure of sleep—add a layer of lived texture, linking daily physiological variation with longer arcs of biological change. When these streams of information are considered together, AI helps form a more integrated portrait of how the body's chemistry changes.

Population-level analyses expand the scope of inquiry into how health shifts across communities. AI systems draw on information from hospitals, wearable devices, and environmental monitors to trace how health trends move through populations. During the COVID-19 pandemic, these platforms were used to study transmission behavior, viral evolution, and pressure on healthcare resources, showing both the reach and the limits of computational modeling. Their use highlighted the need for rigorous validation, transparent oversight, and fair access—reminding researchers that the strength of AI in aging research rests not only on technical design, but on the principles guiding its application.

From Reprogramming to Repair: AI's Role in Regenerative-Medicine

As the body ages, its repair mechanisms change, reflecting shifts in molecular activity, cellular function, and tissue integrity. Different organs respond to stress and accumulated injury in distinct ways, contributing to age-related changes in cartilage, vascular tissue, and neural circuits. Regenerative medicine studies these processes to understand how cells maintain or lose their capacity to restore damaged structures. Artificial intelligence (AI) is becoming part of this work, offering computational tools that model cellular signaling, study how engineered tissues take shape in laboratory systems, and quantify features that contribute to repair.

Stem-cell research shows how AI can refine the study of regenerative change. Guiding stem cells toward defined lineages has long depended on carefully balanced signaling cues, growth factors, and culture conditions. AI deepens this work by analyzing large datasets to identify molecular features most closely associated with successful differentiation, giving researchers clearer insight into how distinct signals shape cellular decisions in laboratory systems. These tools also help reveal how environmental conditions influence developmental pathways and how experimental variability can be reduced, allowing stem-cell models to be examined with greater consistency and precision.

AI is also influencing research on cellular reprogramming. Returning mature cells to more flexible or pluripotent states depends on coordinated changes in transcription factors and epigenetic marks—transitions that machine-learning models can analyze across diverse experimental designs. Partial reprogramming, which examines aging-related change while preserving cellular identity, has become a major area of investigation. Within this work, AI helps researchers adjust reprogramming signals in ways that encourage renewal while maintaining the stability needed for normal cellular function.

In tissue engineering and bioprinting, AI assists in designing scaffolds that present cells with cues for attachment, migration, and maturation. Computational models test how pore geometry, material composition, and spatial arrangement influence the behavior of forming tissues, isolating conditions that support organized cellular structure. Real-time monitoring systems extend this work by allowing AI to track fluctuations in temperature, nutrient flow, and mechanical stress during each stage of tissue development. Although cartilage, skin, and early vascular structures have been constructed, the creation of fully functional organs remains a distant scientific aim rather than an approaching milestone.

Vascularization remains one of regenerative medicine's most persistent hurdles. AI models trained on natural vascular networks allow researchers to follow how endothelial cells extend filaments, select branching points, and stabilize newly formed vessels under shifting biochemical and mechanical cues. These simulations help identify experimental conditions that support more reliable microvessel formation in laboratory systems, offering a clearer view of the early events that make organized vascular growth possible.

AI's reach extends into xenotransplantation and the creation of bio-artificial organs. In xenotransplantation models, machine-learning systems analyze genomic and immunologic data to help researchers examine how donor tissues provoke host immune responses and how targeted genetic modifications alter those reactions. In studies of bio-artificial organs—where engineered materials are joined with living cells—AI helps trace how structural elements, biochemical cues, and cellular activity influence one another at the interface where engineered constructs meet biological tissue. Taken together, these methods give researchers a clearer sense of the conditions under which engineered and biological components can function together.

As regenerative-medicine research grows more demanding, AI offers tools that help scientists organize large datasets, test regenerative ideas through controlled simulations, and design experiments with greater precision. Continued progress will depend on stringent validation, thoughtful integration into experimental workflows, and ethical guidance that keeps pace with technical change. Used in this way, AI can help clarify how aging tissues respond to stress and injury, and how their capacity for repair changes.

Intelligent Machines and Aging Bodies: AI and Robotics in Contemporary Aging

Artificial intelligence and robotics are beginning to influence how aging is studied and experienced, offering new ways to investigate medical care, movement, rehabilitation, and the environments in which daily life unfolds. AI works across large collections of biological, behavioral, and environmental data, giving researchers a clearer view of the factors that shape care and mobility. Robotics provides physical systems through which scientists can study motion, coordination, and human—machine interaction in controlled settings. These technologies widen the scientific and practical questions that can be pursued about aging in a world where digital and mechanical tools are now woven into ordinary life.

Robotic-assisted surgery exemplifies this shift. Systems such as the da Vinci platform enable surgeons to operate using minimally invasive techniques, offering enhanced visualization and mechanical precision. These systems create a stable operative field that makes fine anatomic planes, instrument movement, and the sequence of each surgical step easier for surgeons to see and manage in real time. Although reported outcomes vary across procedures and patient groups, robotic systems are valued for the visibility and mechanical steadiness they provide during complex surgical work.

Robotic systems now play an important role in rehabilitation research. Wearable exoskeletons from Ekso Bionics and ReWalk Robotics allow investigators to track how people with mobility impairments initiate movement, shift weight, and coordinate each step. Robotic gait-training systems such as the Lokomat deliver controlled, feedback-guided cycles of motion that reveal changes in stride, balance, and timing across repeated sessions. When paired with conventional therapy, these devices provide repeatable movement conditions that let clinicians observe how gait and coordination change with continued practice.

AI is also influencing the design of advanced prosthetics. Using machine-learning algorithms and neural interfaces, newer prosthetic limbs can adjust their movements through repeated use, allowing researchers to examine how nerve signals are translated into motion and how control improves with practice. These systems make it possible to measure timing, grip strength, and intended movement with a level of detail that traditional mechanical sensors cannot easily provide. Early work on haptic feedback and artificial proprioception is exploring how engineered limbs might relay sensations such as pressure, joint position, and movement direction back to the user with greater fidelity.

AI and robotics are also finding a place in elder care. Socially assistive robots such as Pepper and PARO can hold short conversations, encourage simple activities, and offer steady points of connection for people who benefit from structured engagement. Intelligent home systems use sensors and machine-learning tools to follow daily routines—how someone moves through their home, prepares meals, or changes sleep and activity habits—and can notify caregivers when something noticeably departs from what is usual for that person. These technologies raise important questions about privacy, independence, and how much of daily care should be entrusted to machines rather than to human caregivers.

Work on nanorobotics remains in its earliest stages, but scientists are beginning to explore how tiny devices might behave inside the body. AI-based models help test how such micro-scale tools could move through bodily fluids, follow chemical cues, or contact specific cellular targets. These studies help investigators consider how biology and engineering might converge at very small scales.

As these systems mature, they do not replace human judgment; rather, they broaden the ways aging can be studied and monitored. The convergence of AI and robotics points to a future in which changes related to aging are easier to notice and interpret, and studied using the sensors, devices, and digital systems already becoming part of daily life.

Mind and Machine: AI and New Approaches to Cognitive and Emotional Research

As life expectancy rises, attention turns toward the cognitive and emotional aspects of growing older. Medicine has extended the number of years people live, but understanding how memory, attention, and emotional experience change across those years remains a demanding scientific task. Artificial intelligence (AI) offers new ways to examine these changes by analyzing information that once required painstaking manual review. AI adds computational depth to the work of understanding how thinking and feeling evolve with age.

A growing area of research focuses on early signs of cognitive change. Machine-learning models trained on brain-imaging and EEG data—and even on speech samples—can identify features that may relate to shifts in attention, memory, or language. Work from multiple research groups suggests that changes in vocabulary, sentence structure, or storytelling style sometimes appear before more obvious cognitive difficulties. These systems offer another way to study how early cognitive changes may emerge.

AI is changing how people access mental-health support. Digital platforms such as Woebot and Wysa use conversational AI to guide users through brief exercises drawn from cognitive behavioral therapy (CBT), including identifying automatic thoughts, labeling emotions, and practicing alternative responses. Because these tools are available at any time, they give people a way to engage with CBT techniques outside traditional clinical settings, without waiting for an appointment or traveling to a clinic.

Adaptive cognitive-training systems are another area of exploration. Programs such as Lumosity, CogniFit, and Elevate adjust task difficulty as users work through memory, attention, or reasoning challenges, creating exercises that respond to a person's performance in real time. Researchers are studying how these platforms relate to mental engagement in older adults and how different tasks place varying demands on attention as difficulty changes.

AI is also beginning to guide research in psychiatry. One area of interest involves pharmacogenomics, a field that uses genetic and biochemical information to examine how a person may process certain medications or show susceptibility to specific side effects. This work remains limited to research settings, because the findings require extensive validation and clinical judgment. Even so, it points toward a future in which clinicians may have clearer information to consider when discussing medication choices with patients.

At a population level, AI can analyze information gathered from social-media posts, online search trends, and wearable devices. These analyses can reveal broad shifts in how communities sleep, move, or express emotion over time. Because they rely on aggregated digital traces, they also raise questions about privacy, fairness, and the appropriate limits of large-scale monitoring.

Taken together, these developments point toward a future in which AI helps illuminate the study of cognitive and emotional aging. AI offers new ways to follow how thinking and feeling change over time. Its value lies not in replacing human insight, but in expanding the depth and scope of the questions that aging research can address.

The Intelligent Future: AI, Aging, and the Promise of a New Era

Artificial intelligence is beginning to identify early biological and behavioral changes that appear long before late-life illness, drawing attention not only to what emerges last, but to what appears earliest. By bringing together information from prediction models, regenerative-biology studies, cognitive tools, and molecular measurements, AI helps reveal how different aspects of aging connect and influence one another.

AI's role in longevity research also carries social implications. Predictive models allow planners and policymakers to explore how longer lives may affect healthcare systems, workplace expectations, and the financial arrangements that support people in later life. A society in which more people reach advanced age will need to consider education, opportunities for continued employment, and ways to sustain economic security across longer lifespans. Taken together, these shifts show that the social structures surrounding aging evolve alongside the science that seeks to understand it.

Even with many questions still unanswered, AI-assisted research is beginning to show how biological processes, daily behaviors, and social settings converge across the course of aging. These developments remind us that later life depends not only on health, but on how people adjust to shifting abilities, relationships, and routines. As AI becomes part of the tools used to study and understand aging, it offers clearer ways to follow the changes that accumulate with time. In doing so, it encourages a more attentive view of growing older and invites reflection on what aging may come to mean as human experience and intelligent technologies develop side by side.

CHAPTER 23

Partnering with Your Healthcare Provider: Building a Longevity Team

"The good physician treats the disease; the great physician treats the patient who has the disease."
—WILLIAM OSLER

In previous chapters, this book has traced the scientific, philosophical, and practical currents that shape contemporary longevity thinking. As we approach the final part of the book, the emphasis shifts—from exploration to orientation. Chapter 23 considers principles that can help readers engage constructively with healthcare professionals, while Chapter 24 offers a structured yet adaptable framework for translating longevity aspirations into coherent plans.

Modern medicine is exceptionally effective in responding to acute crises and immediate threats. Chronic conditions, however, develop gradually and often fall outside the brief, episodic structure of traditional care. Routine checkups and preventive screenings remain important, but they may not detect more subtle physiological patterns—metabolic, inflammatory, or hormonal—that accumulate with age. These features, though easy to miss in daily life, play an important role in long-term physiological change.

Aging follows no single course. Two people with the same clinical measurements may experience very different trajectories because of differences in genetics, metabolism, hormonal profiles, daily habits, and their surrounding environment. Standard guidelines provide useful reference points, but they cannot fully capture the variability inherent in human biology.

Evolving Approaches to Aging and Healthcare

Longevity has always been about more than the number of years lived. It reflects the biological, cognitive, and emotional changes that accompany later life. In turn, modern healthcare is placing greater emphasis on recognizing early patterns in metabolism, inflammation, and hormonal activity, using periodic assessment rather than episodic response to support long-term planning, and accounting for the individual differences that influence clinical care.

Advances in diagnostics and data analysis now allow earlier identification of specific biological changes. Early shifts in metabolism, inflammation, or organ function can prompt more informed conversations between clinicians and patients about long-term priorities. Assessments across major domains—cardiovascular, metabolic, neurological, and musculoskeletal—can then be interpreted alongside daily behavior and environmental factors. When these findings are paired with individualized clinical discussion, clinicians can place a person's health profile within the broader context of aging, clarifying how capacities may develop over time and how lifestyle patterns and clinical care intersect across later decades of life.

A Multidisciplinary View of Aging and Care

Aging research increasingly acknowledges that no single model can account for the complexity of long-term physiological and psychological change. A multidisciplinary approach—integrating conventional medical care with research from multiple scientific disciplines—offers a more complete picture. Collaboration sits at the center of this model. Primary care physicians, longevity-focused clinicians, and allied health professionals work together, each contributing knowledge as a person's needs and priorities evolve.

The Role of Longevity-Focused Clinicians

Longevity-oriented clinicians—whether in preventive, functional, integrative, or geriatric medicine—help interpret the clinical and laboratory findings that accompany aging. Through comprehensive assessments and, when appropriate, advanced testing, they examine measurable changes in metabolism, inflammation, organ function, and hormonal activity, and consider these findings alongside genetics, daily habits, environmental context, and medical history.

They also help people understand complex health information, translating laboratory results and physical measurements into grounded discussions about nutrition, movement, possible supplementation, and, when needed, hormone-related evaluation. They work with patients to develop strategies that fit personal goals and clinical needs. These conversations continue over time as new data emerge, and longevity-focused clinicians often coordinate with primary care providers to maintain continuity of care.

Clinicians also invite people to look closely at daily habits—sleep, stress management, physical activity, and recovery—and how these habits influence the experience of aging. This combined clinical and lifestyle-oriented perspective allows for a clearer view of physical, cognitive, and emotional function across later life.

Forming and Coordinating a Healthcare Team

Forming a healthcare team that aligns with priorities related to aging requires deliberation. Many people begin by considering how their primary care physician approaches prevention, communication, and shared decision-making. When discussions about aging science or long-term planning feel limited, some turn to clinicians who incorporate research on aging biology while remaining firmly rooted in conventional medical care.

Evaluating longevity-oriented or preventive clinicians often starts with practical questions: Do they stay current with developments in aging research? How do they balance standardized guidelines with the details of a person's medical history and daily routines? How do they approach diagnostic decisions and collaboration with other clinicians? These questions invite reflection on whether a clinician's working philosophy aligns with one's own priorities.

Telemedicine, remote consultations, and patient-initiated services have widened access to clinicians who focus on aging-related questions. Working with multiple providers may require active coordination—sharing test results, encouraging communication among clinicians, or, in more complex situations, involving a care coordinator who can help integrate information.

Practical constraints also influence these choices. Insurance coverage may not include longevity-focused visits or specialized testing. When financial considerations arise, people often prioritize services most relevant to their health concerns or look for clinicians who offer different options for engagement. Building a healthcare team becomes an ongoing process—revisited as needs change, new information emerges, and goals are clarified.

A Foundation of Trust and Flexibility

A durable healthcare partnership rests on trust, curiosity, and a willingness to adapt. Clinicians who stay attentive to developments in aging science and communicate with clarity create a setting where priorities can be explored openly and with care. As goals change and life circumstances shift, care plans benefit from a shared readiness to adjust, allowing decisions to remain aligned with what matters most at each stage.

These relationships offer a steady base from which long-term decisions can be approached with greater insight. When care is anchored in honest communication, genuine collaboration, and thoughtful reflection, the conditions are created to meet the later years with a calmer perspective and a clearer sense of place in the world.

CHAPTER 24

Longevity by Design: Building a Personal Strategy

"A goal without a plan is just a wish."
—ANTOINE DE SAINT-EXUPÉRY

Two people may share an aspiration to approach aging with care and foresight, yet their paths may diverge sharply. One proceeds deliberately, guided by thoughtful, adaptable goals grounded in scientific understanding. The other moves restlessly from one idea to another, experimenting with diets, supplements, and routines without a coherent vision. Access to knowledge may be the same, but only the person who plans with intention and revisits those choices consistently tends to translate information into practices that endure.

A well-crafted longevity plan offers more than structure; it brings clarity amid complexity. In a landscape shaped by rapid scientific advances and shifting cultural trends, a personalized roadmap helps filter noise, highlight meaningful choices, and orient routines toward principles with durable grounding. Without such intentionality, it is easy to slip into a reactive stance, responding to concerns as they arise rather than situating them within a larger narrative. Purposeful decisions made over time cultivate steadier direction.

There is no singular approach to aging. Research reveals the influence of genetics, metabolism, behavior, and environment, yet people respond differently to the same practices. Intermittent fasting may align well with one person's biological rhythms while proving misaligned for another. Supplements may complement one physiology and feel unnecessary or counterproductive for someone else. Even something as widely valued as exercise requires attention to history, capability, and circumstance.

Longevity develops not through isolated efforts but through habits that reinforce one another. Nutrition, movement, sleep, and other foundational practices influence the body in interconnected ways. A longevity plan guided by this understanding favors sustainability over short-lived trends and balance over extremes.

The following sections provide tools for assessing current conditions, clarifying priorities, and integrating insights into a plan that remains adaptive and open to revision. Aging proceeds regardless, but the habits cultivated—physical, cognitive, and emotional—can give direction and continuity to how those years are lived.

Know Thyself: Establishing a Biological Foundation for Long-Term Planning

Every serious longevity effort begins with translation: moving from intention to action by identifying a biological starting point. A thoughtful approach rests on an honest, multidimensional view of physiology. Routine medical screenings—lipid panels, fasting glucose, blood pressure, and standard bloodwork—form the foundation of that view. Longevity-oriented assessments widen the scope further, incorporating immune markers, hormonal patterns, genetic predispositions, and measures linked to cellular aging.

More advanced diagnostics can offer additional insight into metabolic and inflammatory activity. Tests such as homocysteine and C-reactive protein (CRP) help clarify how the body responds to physiological stress. Cardiovascular markers—fasting insulin, triglyceride-to-HDL ratios—and routine assessments of liver and kidney function provide further information about key organ systems. Hormone evaluations, including testosterone, estrogen, thyroid measures, and cortisol levels, can help explain changes in energy, mood, and overall physiology.

For those interested in contemporary aging science, biological-age assessments introduce an additional perspective. Epigenetic clocks, which examine DNA methylation patterns, offer information about how environmental and lifestyle factors relate to cellular activity. Telomere-length measurements provide another indicator associated with cellular aging. Some people also consider tests of mitochondrial function to examine how efficiently cells produce energy and how these patterns correspond with broader physiological function.

Physical and cognitive assessments add valuable detail to a longevity evaluation. Grip strength, gait speed, muscle mass, bone density, and cardiovascular performance provide measurable data on physical capacity. Cognitive tests that examine memory, reaction time, and processing speed offer specific information about aspects of neurological function. Data on sleep, emotional state, and physiological stress responses contribute additional context about day-to-day functioning.

Modern tracking technologies—wearables, continuous glucose monitors, smartwatches, and biometric rings—supply real-time physiological data. They can show changes in activity, sleep, heart rate, and glucose levels that may be considered alongside clinical and laboratory findings.

Establishing an adaptive baseline provides a clear starting point for observing change over time. It supports a practical approach to planning—one that remains informed, contextual, and responsive as new information emerges.

From Insight to Action: Interpreting a Personal Health Profile

Once a biological baseline has been established, the task becomes interpretation—using the information to understand current position and direction. Biological, cognitive, and emotional assessments outline present status in concrete terms, making it easier to evaluate how different choices may influence the years ahead.

The first step is identifying what the findings reveal. Some results may warrant closer attention—such as markers related to inflammation, glucose regulation, or hormonal activity—while others may point to areas of relative strength, including cardiovascular fitness or muscular capacity. Daily routines can then be examined in relation to stated aims by looking closely at nutrition, physical activity, sleep, and stress-management practices. Genetic background and family history add further context, guiding which questions deserve closer scrutiny.

Mental and emotional factors merit the same level of consideration. Sources of strain or sustained stress may signal areas where additional support is needed, while periods of calm, focus, or adaptability offer insight into existing resilience and coping capacity.

At this analytical stage, feasibility becomes central. Rather than pursuing broad or sweeping change, attention can be directed toward specific capacities to strengthen—such as stamina, metabolic stability, or cognitive focus—and toward approaches that remain workable over time. Structured goal-setting frameworks, often used in planning and behavioral research, can lend clarity to aims that might otherwise remain diffuse.

From Aspirations to Achievement: Turning Health Priorities into Practical Steps

Moving from intention to action requires both clarity and adaptability. Primary aims can be identified—whether related to cardiovascular, metabolic, neurological, or emotional health—and distinguished from secondary interests such as steadier sleep or more effective stress management. A focus on metabolic questions may lead to monitoring glucose patterns, while an interest in cognitive function may prompt periodic evaluation of memory or overall cognitive performance.

Whatever the emphasis, a consistent method for observing progress supports continuity—whether through a written journal, a digital log, or scheduled conversations with healthcare professionals. Such practices create a responsive feedback loop, allowing plans to evolve as scientific understanding grows and life circumstances shift. In this way, aspirations translate into approaches that remain practical and responsive to lived experience.

Blueprint for Lifelong Health: Building a Maintainable, Adaptive Lifestyle Plan

Once goals are defined, the next step is implementation—establishing routines that can be maintained over time. Core practices often form the basis of a long-term plan: eating in ways that support stable energy and adequate nutrition, exercising regularly to build strength and cardiovascular capacity, maintaining mobility, and following sleep routines that aid recovery. These are practical elements that help structure daily life.

More specific methods may be added as circumstances require. Intermittent fasting adjusts meal timing with the aim of influencing metabolic patterns. High-intensity interval training varies exercise intensity to challenge cardiovascular and respiratory systems. Sauna use and cold exposure alter temperature conditions in ways that affect heart rate, circulation, and perceived stress. Pharmacological or nutraceutical approaches—particularly compounds discussed in aging research—require caution, professional oversight, and careful consideration of health context.

A Wellness Team

Implementing a long-term plan is rarely a solitary endeavor. As discussed in Chapter 23, strong relationships with healthcare professionals—built on communication, curiosity, and shared purpose—create a framework in which long-term plans can adjust as new information emerges or life circumstances change.

Balance Over Perfection: Navigating Common Challenges

Perfectionism is an obstacle in any long-term effort. The urge to do everything "right" can create pressure that disrupts consistency. Instead of aiming for rigid precision, many people find it more practical to incorporate science-informed habits into daily routines—practices that feel workable rather than forced.

Data can be useful, but excessive monitoring can add stress rather than insight. Balancing objective measurements with physical and emotional cues allows for clearer judgment and steadier decision-making. Staying connected to others, managing emotions with flexibility, and taking part in valued activities all contribute to a more stable foundation for daily choices.

A sustainable plan draws on scientific understanding and self-compassion. It acknowledges that progress varies from day to day, yet steady intention and workable habits help support a more deliberate and manageable experience of aging.

The Feedback Loop: Monitoring and Refining a Health Strategy

A thoughtful health strategy does not remain fixed. It shifts as circumstances change—guided by emerging research, evolving responsibilities, and the unfolding patterns of daily life. Once a baseline and early aims are established, returning to them periodically becomes a way of staying oriented as the years move forward.

Regular check-ins—whether every six months, annually, or after major life changes—create space to review what has been gathered. Biomarkers, physical measurements, cognitive tests, and everyday observations often take on meaning only when viewed across longer intervals. Wearables and routine tracking extend this perspective by showing gradual changes in activity, sleep, heart rate, and other physiological measures.

Bringing objective findings together with professional insight and lived experience allows plans to adjust with intention rather than urgency, creating a steady process of recalibration over time. In this way, longevity planning maintains continuity—remaining flexible as circumstances change while staying grounded in enduring priorities and leaving room for growth.

Bringing It All Together: Creating a Sustainable Longevity Plan

With a clearer sense of biological context, priorities, and the strategies that support them, attention turns to building a plan that can be carried into daily life. A practical plan fits the contours of everyday experience—schedule, temperament, responsibilities, and long-term aims. Its strength lies in its capacity to be lived rather than merely imagined.

A longevity-oriented plan becomes evident not only in data but in the details of daily routine: more stable habits, a clearer sense of direction, and choices guided by values such as curiosity, autonomy, steadiness, and meaningful connection. Consistency plays a central role. The aim is not rigid adherence, but a structure that can adjust as circumstances and understanding shift.

A workable approach often begins with a daily rhythm aligned with available time and capacity. Periodic checkpoints—brief reflections within the day, more focused reviews across weeks, and broader evaluations over longer intervals—create opportunities to revisit priorities and recalibrate direction. A written plan provides continuity, evolving as knowledge deepens and life circumstances change.

Over time, deliberate planning becomes a steady discipline. Longevity shifts from a measure of years to a way of living shaped by sustained priorities and thoughtful adjustment.

Aging with Intention: Designing a Purpose-Driven Life

Conscious living begins where knowledge, habit, and personal meaning meet. Science can outline the terrain, but the experience of moving through it develops only when daily practices take hold—when nourishing meals, regular movement, restorative sleep, and emotional steadiness become part of lived experience.

As these routines settle, they create space for refinement: engagement with new research, consideration of emerging technologies in conversation with qualified professionals, and adjustment of habits that once served well but now require change. A durable plan shifts with circumstance, balancing curiosity with sustained attention to fundamental practices.

The aim extends beyond longevity. Conscious living brings physical experience, emotional balance, intellectual engagement, and sustained participation into closer alignment. Foundational habits support clearer decisions, and ongoing inquiry becomes part of daily life rather than a separate pursuit.

Emerging developments—regenerative medicine, genetic research, new technologies, and artificial intelligence—will continue to broaden understanding of aging. Working with healthcare professionals helps ensure new insights are integrated thoughtfully rather than reactively. In this way, mindful living moves longevity from a measure of time to a way of approaching life marked by stability, connection, and continued learning.

In the end, aging with intention reflects not a wish to escape time, but the courage to live exposed to it.

> *"Life is never made unbearable by circumstances,*
> *but only by lack of meaning and purpose."*
> —VIKTOR FRANKL

Postscript

The Age of Possibility: A Final Reflection

Aging was once regarded as an unavoidable force—something to endure, resist, or delay. Today, it is understood differently. Aging is not a single event but a continuous process shaped by biology, time, and daily choices.

Historically, longevity depended on genetics, environment, and the limits of medical knowledge. Research now offers a clearer view of how aging progresses within the body. Scientists can track how metabolism changes, how inflammatory signals fluctuate, and how cellular repair becomes less efficient over time. Studies in brain science and physical performance shed light on shifts in thinking, memory, and strength. These findings cannot predict any one person's experience, but they help clarify the forces that influence the aging process.

Longevity begins in science and ends in philosophy. A longer life gains meaning when guided by purpose, curiosity, and deliberate engagement. Its value lies not in the number of years lived but in the depth found within them—in the ideas pursued, relationships sustained, and moments that endure.

Throughout this book, research related to physical, cognitive, and emotional function has been examined. Yet reading about longevity is not the same as understanding what supports a longer life. Knowledge becomes useful only when carried into daily behavior. A longevity plan grows from consistent habits and a clear intention that guides decisions over time.

Science continues to refine what is known about aging, yet some truths remain. No drug, genetic intervention, or scientific advance can replace purpose or human connection. Longevity is neither a count of years nor a retreat from mortality; it is a way of meeting each day with attentiveness and gratitude, aware that time still allows room for growth and deeper engagement with life.

Perhaps the most important recognition is that aging is not passive. We influence how it unfolds. Meals prepared with care, movement that keeps the body active, rest that restores, and conversations that matter all contribute to the rhythm of daily life. In these ordinary moments, the foundation of later years is laid.

As understanding of aging shifts, its influence reaches far beyond personal experience. The pursuit of longevity now touches how work, community, relationships, and purpose are imagined. When the idea of inevitable decline is set aside, new possibilities come into view. And as science advances, one question endures: how will time be used?

This question remains open. The future of aging is not predetermined, but shaped—gradually, deliberately—by the choices that give life direction and meaning.

This book was written in the belief that aging deserves attention rather than denial, curiosity rather than fear, and care rather than urgency. If it has clarified even a small part of how time is met, it has done what we set out to do.

Drs. Ioulia and Don Howard

> *"What we call the beginning is often the end.*
> *And to make an end is to make a beginning."*
> —T. S. Eliot

References

About These References

The references included in this section provide the scientific foundation for the strategies, research, and concepts explored throughout this book. They encompass historical and philosophical perspectives on aging as well as contemporary developments in longevity science, supporting discussions of physical activity, nutrition, supplements, fasting, sleep, stress management, emotional well-being, sex, hormonal health, pharmacological approaches, biohacking, regenerative medicine, gene therapy, artificial intelligence, and personalized longevity planning.

To enhance usability, references are organized by subject rather than presented in a single alphabetical list, allowing readers to locate material relevant to specific topics efficiently. Within each category, references are arranged alphabetically by the first author's last name and formatted according to APA (7th edition) guidelines. Digital object identifiers (DOIs) or URLs are included where available to facilitate access to original research articles, authoritative reports, and key texts.

Longevity science is expansive and continues to evolve. The aim of this book has been to distill that complexity into clear, engaging, and practically relevant material—balancing depth with readability and emphasizing conceptual understanding rather than exhaustive technical detail. Accordingly, the level of detail is intended for an intelligent, scientifically curious readership rather than for specialist audiences. Readers who wish to follow the field in greater depth are encouraged to engage directly with the peer-reviewed literature and other authoritative sources as research continues to advance.

History of Aging

Achenbaum, W. A. (2013). A historical perspective in aging and gerontology. In D. Dannefer & C. Phillipson (Eds.), The SAGE handbook of social gerontology (pp. 3–13). SAGE Publications. https://doi.org/10.4135/9781446200933.n2

Center for Healthy Aging. (2022, February 21). History of aging research. Colorado State University. https://www.research.colostate.edu/healthyagingcenter/2022/02/21/history-of-aging-research/

Chappell, J. (2024). Golden years: How Americans invented and reinvented old age. Atlantic Books.

Dowling, M., Kenney, C., & Carney, M. T. (2024). The aging revolution: The history of geriatric health care and what really matters to older adults. Northwell Health Press. https://www.johnahartford.org/dissemination-center/view/the-aging-revolution-the-history-of-geriatric-health-care-and-what-really-matters-to-older-adults-northwell-health

Haber, C. (2011). A cultural and economic history of old age in America. Mayo Clinic Proceedings, 86(9), 848–850. https://doi.org/10.4065/mcp.2011.0265

Katz, S. (2018). Aging through the lens of historical time, space, and place. The Gerontologist, 58(4), 795–802. https://doi.org/10.1093/geront/gnx195

University of Cambridge. (2010, May 24). A brief history of ageing. University of Cambridge News. https://www.cam.ac.uk/research/news/a-brief-history-of-ageing

Philosophy and Aging

Aristotle. (2019). *Nicomachean ethics*. Cambridge University Press.

Frankl, V. E. (2006). *Man's search for meaning*. Beacon Press.

Marcus Aurelius. (2002). *Meditations* (G. Hays, Trans.). Modern Library.

Nietzsche, F. (1992). *Ecce homo*. Penguin Classics.

Seneca. (2004). *On the shortness of life*. Penguin Classics.

Buettner, D. (2012). *The Blue Zones: Lessons for living longer from the people who've lived the longest*. National Geographic.

Davidson, R. J., & McEwen, B. S. (2012). Social influences on neuroplasticity: Stress and interventions to promote well-being. *Nature Neuroscience, 15*(5), 689–695. https://doi.org/10.1038/nn.3093

Ryff, C. D., & Singer, B. H. (1998). The contours of positive human health. *Psychological Inquiry, 9*(1), 1–28. https://doi.org/10.1207/s15327965pli0901_1

Waldinger, R., & Schulz, M. (2023). *The good life: Lessons from the world's longest scientific study of happiness*. Simon & Schuster.

Weil, A. (2005). *Healthy aging: A lifelong guide to your well-being*. Knopf.

García, H., & Miralles, F. (2017). *Ikigai: The Japanese secret to a long and happy life*. Penguin Books.

Haidt, J. (2006). *The happiness hypothesis: Finding modern truth in ancient wisdom*. Basic Books.

Purpose and Longevity

Epel, E. S., Daubenmier, J., Moskowitz, J. T., Folkman, S., & Blackburn, E. (2009). Meditation practice is associated with longer telomeres in women. *Annals of the New York Academy of Sciences, 1172*, 34–53. https://doi.org/10.1196/annals.1393.001

Hill, P. L., & Turiano, N. A. (2014). Purpose in life as a predictor of mortality across adulthood. *Psychological Science, 25*(7), 1482–1486. https://doi.org/10.1177/0956797614531799

Holt-Lunstad, J., Smith, T. B., & Layton, J. B. (2010). Social relationships and mortality risk: A meta-analytic review. *PLOS Medicine, 7*(7), e1000316. https://doi.org/10.1371/journal.pmed.1000316

Sapolsky, R. M. (2004). *Why zebras don't get ulcers*. Holt Paperbacks.

Biological Theories of Aging

Aquino, T., & Leuzzi, L. (2025). Finite-size effects in aging can be interpreted as sub-aging. *arXiv*. https://arxiv.org/abs/2501.04843

Gems, D., & de Magalhães, J. P. (2021). The hoverfly and the wasp: A critique of the hallmarks of aging as a paradigm. *Ageing Research Reviews, 70*, Article 101407. https://doi.org/10.1016/j.arr.2021.101407

Kennedy, B. K., Berger, S. L., Brunet, A., Campisi, J., Cuervo, A. M., Epel, E. S., Franceschi, C., Lithgow, G. J., Morimoto, R. I., Pessin, J. E., Rando, T. A., Richardson, A., Schadt, E. E., Wyss-Coray, T., & Sierra, F. (2014). Geroscience: Linking aging to chronic disease. *Cell, 159*(4), 709–713. https://doi.org/10.1016/j.cell.2014.10.039

Lemaître, J. F., Ronget, V., & Gaillard, J. M. (2024). The evolution of ageing: Classic theories and emerging ideas. *Biogerontology, 25*(1), 1–15. https://doi.org/10.1007/s10522-024-10143-5

López-Otín, C., Blasco, M. A., Partridge, L., Serrano, M., & Kroemer, G. (2013). The hallmarks of aging. *Cell, 153*(6), 1194–1217. https://doi.org/10.1016/j.cell.2013.05.039

López-Otín, C., Blasco, M. A., Partridge, L., Serrano, M., & Kroemer, G. (2023). Hallmarks of aging: An expanding universe. *Cell, 186*(2), 243–278. https://doi.org/10.1016/j.cell.2022.12.038

Maklakov, A. A., & Regan, J. C. (2025). Consolidating multiple evolutionary theories of ageing suggests a need for new approaches to study genetic contributions to ageing decline. *Ageing Research Reviews*. Advance online publication. https://doi.org/10.1016/j.arr.2024.102201

Medawar, P. B. (1952). *An unsolved problem of biology*. H. K. Lewis.

Pridham, G., & Rutenberg, A. D. (2023). Dynamical network stability analysis of multiple biological ages provides a framework for understanding the aging process. *arXiv*. https://arxiv.org/abs/2309.10005

Qin, H. (2024). The emergent aging model: Aging as an emergent property of biological systems. *arXiv*. https://arxiv.org/abs/2407.05226

Tapias, D., Marteau, C., Aguirre-López, F., & Sollich, P. (2024). Bringing together two paradigms of non-equilibrium: Driven dynamics of aging systems. *arXiv*. https://arxiv.org/abs/2402.03516

The hallmarks of aging as a conceptual framework for health and disease. (2024). *Frontiers in Aging*. https://doi.org/10.3389/fragi.2024.1334261

Longevity.Technology. (2024). *Theories of aging: Advancing understanding in 2024*. https://longevity.technology/news/theories-of-aging-advancing-understanding-in-2024/

Viña, J., Borrás, C., & Miquel, J. (2007). Theories of ageing. *IUBMB Life*, *59*(4–5), 249–254. https://doi.org/10.1080/15216540701226011

Wang, Y., & Zhu, L. (2025). Molecule-dynamic-based aging clock and aging roadmap forecast with Sundial. *arXiv*. https://arxiv.org/abs/2501.02176

Wang, Z., Chen, Y., Li, X., Liu, H., & Zhang, Q. (2024). Pro-aging metabolic reprogramming: A unified theory of aging. *Engineering*. Advance online publication. https://doi.org/10.1016/j.eng.2024.09.010

The Epigenetic Clock as a Measure of Biological Aging

Brigham and Women's Hospital. (2024, February 14). New epigenetic clocks reinvent how we measure age. *ScienceDaily*. https://www.sciencedaily.com/releases/2024/02/240214203341.htm

CheekAge. (2024). A next-generation epigenetic buccal clock predictive of mortality in human blood. *Frontiers in Aging*. https://doi.org/10.3389/fnagi.2024.01023

Garma, L., & Quintela-Fandino, M. (2024). Applicability of epigenetic age models to next-generation methylation arrays. *Genome Medicine*, *16*(1), Article 116. https://doi.org/10.1186/s13073-024-0116-0

Leroy, A., Teh, A. L., Dondelinger, F., Alvarez, M. A., & Wang, D. (2023). Longitudinal prediction of DNA methylation to forecast epigenetic outcomes. *arXiv*. https://doi.org/10.48550/arXiv.2312.13302

Voisin, S., Harvey, N. R., Haupt, L. M., Griffiths, L. R., Ashton, K. J., & Coffey, V. G. (2023). Influence of physical activity on the epigenetic clock: Evidence from a monozygotic twin study. *Clinical Epigenetics*, *15*, Article 56. https://doi.org/10.1186/s13148-023-01756-4

Psychological Theories of Aging

Min, J., Yoo, H. J., Nashiro, K., & Mather, M. (2023). Modulating heart rate oscillation affects plasma amyloid beta and tau levels in younger and older adults. *Scientific Reports, 13,* Article 3967. https://doi.org/10.1038/s41598-023-30167-0

Nashiro, K., Yoo, H. J., Min, J., & Mather, M. (2024). Increasing coordination and responsivity of emotion-related brain regions with a heart rate variability biofeedback randomized trial. *Cognitive, Affective, & Behavioral Neuroscience.* Advance online publication. https://doi.org/10.3758/s13415-023-01100-7

Shenkman, G., Ifrah, K., & Shmotkin, D. (2023). The contribution of couplehood and parenthood to the hedonic and eudaimonic well-being of older gay men. *Journal of Happiness Studies, 24,* 1234–1250. https://doi.org/10.1007/s10902-022-00567-8

Stine-Morrow, E. A. L., West, R., Cheng, S., Abrams, L., Gerstorf, D., Hooker, K., Kunzmann, U., & Lustig, C. (2024). Advancing theory-driven research in the psychological science of adult development and aging. *Psychology and Aging, 39*(8). https://doi.org/10.1037/pag0000865

Sociological Theories of Aging

Katz, S. (2024). Sociology and gerontology: New perspectives and vital issues. *Innovation in Aging, 8*(Supplement 1), 601. https://doi.org/10.1093/geroni/igy023.601

Katz, S., & Marshall, B. L. (2024). New materialist insights in the sociology of aging: Rethinking agency and context. *Innovation in Aging, 8*(Supplement 1), 602. https://doi.org/10.1093/geroni/igy023.602

Sharma, M., & Sharma, S. (2023). Sociological theories of ageing and their influence on the quality of life of older adults. *Research and Reviews: Journal of Geriatric Nursing and Health Sciences, 9*(3), 17–22. https://matjournals.net/nursing/index.php/RRJGNHS/article/view/17

Shmotkin, D. (2023). Psychosocial and biological pathways to aging. *Zeitschrift für Gerontologie und Geriatrie, 56*(5), 482–489. https://doi.org/10.1007/s00391-024-02324-1

Exercise

Ahern, S., Sharma, R., & Mathes, D. (2025). Importance of exercise for career longevity, injury prevention, and mental health among plastic surgeons. *Plastic and Reconstructive Surgery—Global Open, 13*(1), e5256. https://doi.org/10.1097/GOX.0000000000005256

Andersen, K., Pedersen, B. K., Kujala, U. M., & Singh, M. A. F. (2025). Global exercise recommendations for healthy ageing: A multifaceted approach for older adults. *Age and Ageing, 54*(1), Article afad001. https://doi.org/10.1093/ageing/afad001

Booth, F. W., Roberts, C. K., & Laye, M. J. (2012). Lack of exercise is a major cause of chronic diseases. *Comprehensive Physiology, 2*(2), 1143–1211. https://doi.org/10.1002/cphy.c110025

Conroy, G. (2024, May). Why is exercise good for you? Scientists are finding answers in our cells. *Nature.* https://www.nature.com/articles/d41586-024-01200-7

D'Onofrio, G., Kirschner, J., Prather, H., Goldman, D., & Rozanski, A. (2023). Musculoskeletal exercise: Its role in promoting health and longevity. *Progress in Cardiovascular Diseases, 77*, 25–36. https://doi.org/10.1016/j.pcad.2023.02.006

Falshaw, N., Sagner, M., & Siow, R. C. (2024). The Longevity Med Summit: Insights on healthspan from cell to society. *Frontiers in Aging, 5*, Article 1417455. https://doi.org/10.3389/fragi.2024.1417455

Garber, C. E., Blissmer, B., Deschenes, M. R., Franklin, B. A., Lamonte, M. J., Lee, I.-M., Nieman, D. C., & Swain, D. P. (2011). American College of Sports Medicine position stand: Quantity and quality of exercise for developing and maintaining cardiorespiratory, musculoskeletal, and neuromotor fitness in apparently healthy adults. *Medicine & Science in Sports & Exercise, 43*(7), 1334–1359. https://doi.org/10.1249/MSS.0b013e318213fefb

Guan, Y., & Yan, Z. (2022). Molecular mechanisms of exercise and healthspan. *Cells, 11*(5), Article 872. https://doi.org/10.3390/cells11050872

Hamilton, K. L., & Selman, C. (2023). Can exercise prevent the age-related decline in adaptive homeostasis? Evidence across organisms and tissues. *The Journal of Physiology.* https://doi.org/10.1113/JP284583

Harvard Health Publishing. (2023). Adding weight lifting to workouts may boost longevity. *Harvard Health.* https://www.health.harvard.edu/heart-health/adding-weight-lifting-to-workouts-may-boost-longevity

Harvard Health Publishing. (2023). Strength training might lengthen life. *Harvard Health.* https://www.health.harvard.edu/staying-healthy/strength-training-might-lengthen-life

Harvard T.H. Chan School of Public Health. (2023). Exercising more than recommended could lengthen life, study suggests. *Harvard T.H. Chan School of Public Health News.* https://www.hsph.harvard.edu/news/hsph-in-the-news/exercising-more-than-recommended-could-lengthen-life-study-suggests/

Khan, M., Al Saud, H., Sierra, F., Perez, V., Greene, W., Al Asiry, S., Pathai, S., & Torres, M. (2024). Global Healthspan Summit 2023: Closing the gap between healthspan and lifespan. *Nature Aging, 4*(4), 445–448. https://doi.org/10.1038/s43587-024-00593-4

Lee, D. H., Rezende, L. F. M., Joh, H.-K., Keum, N., Ferrari, G., Rey-Lopez, J. P., Rimm, E. B., Tabung, F. K., & Giovannucci, E. L. (2022). Long-term leisure-time physical activity intensity and all-cause and cause-specific mortality: A prospective cohort of US adults. *Circulation, 146*(7). https://doi.org/10.1161/CIRCULATIONAHA.121.058162

McKee, A., Ortega, F. B., & Lavie, C. J. (2021). Exercise and lifespan: A genomic and epigenomic perspective. *Progress in Cardiovascular Diseases, 64*, 65–72. https://doi.org/10.1016/j.pcad.2020.12.002

Powell, K. E., King, A. C., Buchner, D. M., Campbell, W. W., DiPietro, L., Erickson, K. I., Hillman, C. H., Jakicic, J. M., Janz, K. F., Katzmarzyk, P. T., Kraus, W. E., Macko, R. F., Marquez, D. X., McTiernan, A., Pate, R. R., Pescatello, L. S., Whitt-Glover, M. C., & American College of Sports Medicine. (2019). The scientific foundation for the Physical Activity Guidelines for Americans, 2nd edition. *Journal of Physical Activity and Health, 16*(1), 1–11. https://doi.org/10.1123/jpah.2018-0618

Ruegsegger, G. N., & Booth, F. W. (2018). Health benefits of exercise. *Cold Spring Harbor Perspectives in Medicine, 8*(7), Article a029694. https://doi.org/10.1101/cshperspect.a029694

Seals, D. R., Justice, J. N., & LaRocca, T. J. (2016). Physiological geroscience: Targeting function to increase healthspan and lifespan. *The Journal of Physiology, 594*(8), 2001–2024. https://doi.org/10.1113/JP270538

Swainson, M. G., Biddle, S. J. H., & Stamatakis, E. (2025). Physical activity intensity versus volume: Which matters more for reducing mortality risk? *European Journal of Preventive Cardiology, 32*(1), 10–11. https://doi.org/10.1093/eurjpc/zvad121

Vieira, R. F. L., Junqueira, R. L., Gaspar, R. C., Munoz, V. R., & Pauli, J. R. (2023). Exercise activates AMPK signaling: Impact on glucose uptake in the skeletal muscle in aging. *Rehabilitation Journal*. https://www.rehabiljournal.com/articles/exercise-activates-ampk-signaling-impact-on-glucose-uptake-in-the-skeletal-muscle-in-aging.html

Warburton, D. E., & Bredin, S. S. (2017). Health benefits of physical activity: A systematic review of current systematic reviews. *Current Opinion in Cardiology, 32*(5), 541–556.

Additional News & Industry Sources

American Council on Exercise. (2023). *Exercise and longevity: Empowering and motivating clients*. https://www.acefitness.org/resources/pros/expert-articles/8752/exercise-and-longevity-empowering-and-motivating-clients/

American Council on Exercise. (2023). *Longevity and exercise: Understanding lifespan vs. healthspan*. https://www.acefitness.org/resources/pros/expert-articles/8730/the-longevity-buzzword-understanding-aging-lifespan-healthspan-and-the-role-of-exercise/

American Medical Association. (2023). *Massive study uncovers how much exercise is needed to live longer*. American Medical Association News. https://www.ama-assn.org/delivering-care/public-health/massive-study-uncovers-how-much-exercise-needed-live-longer

GQ. (2023). *What's the best workout for longevity?* GQ Magazine. https://www.gq.com/story/best-workout-for-longevity

Legendary Strength. (2023). *Surprising longevity insights from a new fitness study*. Legendary Strength Blog. https://legendarystrength.com/surprising-longevity-insights-from-a-new-fitness-study/

OnePeloton. (2023). *Exercise for longevity: Workout routines for a lengthy life*. Peloton Blog. https://www.onepeloton.com/blog/exercise-for-longevity/

Nutrition and Diet

Bajerska, J., Chmurzynska, A., Muzsik, A., Krzyzanowska, P., Madry, E., Malinowska, A. M., & Walkowiak, J. (2018). Weight loss and metabolic health effects from energy-restricted Mediterranean and Central-European diets in postmenopausal women: A randomized controlled trial. *Scientific Reports, 8*, Article 11170. https://doi.org/10.1038/s41598-018-29495-3

de Cabo, R., & Mattson, M. P. (2019). Effects of intermittent fasting on health, aging, and disease. *The New England Journal of Medicine, 381*(26), 2541–2551. https://doi.org/10.1056/NEJMra1905136

Dong, L., Teh, D. B. L., Kennedy, B. K., et al. (2023). Unraveling female reproductive senescence to enhance healthy longevity. *Cell Research, 33*(1), 11–29. https://doi.org/10.1038/s41422-022-00718-7

Estruch, R., Ros, E., Salas-Salvadó, J., Covas, M. I., Corella, D., Arós, F., & Martínez-González, M. A. (2023). Primary prevention of cardiovascular disease with a Mediterranean diet supplemented with extra-virgin olive oil or nuts. *The New England Journal of Medicine, 390*(3), 390–401. https://doi.org/10.1056/NEJMoa1800389

Estruch, R., Ros, E., Salas-Salvadó, J., et al. (2018). Primary prevention of cardiovascular disease with a Mediterranean diet supplemented with extra-virgin olive oil or nuts. *The New England Journal of Medicine, 378*(25), e34.

Fontana, L., & Partridge, L. (2023). Promoting health and longevity through diet: From model organisms to humans. *Cell, 184*(3), 1–17. https://doi.org/10.1016/j.cell.2023.03.002

Fraser, G. E., & Shavlik, D. J. (2001). Ten years of life: Is it a matter of choice? *Archives of Internal Medicine, 161*(13), 1645–1652. https://doi.org/10.1001/archinte.161.13.1645

Greenhill, C. (2024). The complex effects of dietary restriction on longevity and health. *Nature Reviews Endocrinology, 20*(10), 697. https://doi.org/10.1038/s41574-024-01051-2

Hu, F. B. (2023). Diet strategies for promoting healthy aging and longevity: An epidemiological perspective. *Journal of Internal Medicine.* https://doi.org/10.1111/joim.13728

Hu, F. B., et al. (2025). Healthy eating in midlife linked to overall healthy aging. *Nature Medicine.* https://doi.org/10.1038/s41591-025-01234-5

Jacquier, E. F., Kassis, A., Marcu, D., Contractor, N., Hong, J., Hu, C., Kuehn, M., Lenderink, C., & Rajgopal, A. (2024). Phytonutrients in the promotion of healthspan: A new perspective. *Frontiers in Nutrition, 11.* https://doi.org/10.3389/fnut.2024.1409339

Levine, M. E., Suarez, J. A., Brandhorst, S., Balasubramanian, P., Cheng, C.-W., Madia, F., & Longo, V. D. (2014). Low protein intake is associated with a major reduction in IGF-1, cancer, and overall mortality in the 65 and younger but not older population. *Cell Metabolism, 19*(3), 407–417. https://doi.org/10.1016/j.cmet.2014.02.006

Longo, V. D., & Anderson, R. M. (2023). Nutrition, longevity and disease: From molecular mechanisms to interventions. *Cell, 185*(9), 145–156. https://doi.org/10.1016/j.cell.2023.09.022

MacArthur, M. R., & Mitchell, S. J. (2023). Sex differences in healthspan and lifespan responses to geroprotective dietary interventions in preclinical models. *Current Opinion in Physiology, 33,* 100651. https://doi.org/10.1016/j.cophys.2023.100651

Mattison, J. A., Colman, R. J., Beasley, T. M., Allison, D. B., & Anderson, R. M. (2017). Caloric restriction improves health and survival of rhesus monkeys. *Nature Communications, 8,* 14063. https://doi.org/10.1038/ncomms14063

Mishra, A., Giuliani, G., & Longo, V. D. (2024). Nutrition and dietary restrictions in cancer prevention. *Biochimica et Biophysica Acta—Reviews on Cancer, 1879*(1), 189063. https://doi.org/10.1016/j.bbcan.2023.189063

Morris, M. C., Tangney, C. C., Wang, Y., Sacks, F. M., Barnes, L. L., Bennett, D. A., & Aggarwal, N. T. (2015). MIND diet associated with reduced incidence of Alzheimer's disease. *Alzheimer's & Dementia, 11*(9), 1007–1014. https://doi.org/10.1016/j.jalz.2014.11.009

Morze, J., Danielewicz, A., Przybyłowicz, K., Zeng, H., Hoffmann, G., & Schwingshackl, L. (2021). An updated systematic review and meta-analysis on adherence to Mediterranean diet and risk of cancer. *European Journal of Nutrition, 60*(3), 1561–1586. https://doi.org/10.1007/s00394-020-02346-6

Park, S.-H., Lee, D.-H., & Jung, C. H. (2024). Scientific evidence of foods that improve the lifespan and healthspan of different organisms. *Nutrition Research Reviews, 37*(1), 169–178. https://doi.org/10.1017/S0954422423000136

Schwingshackl, L., & Hoffmann, G. (2014). Adherence to Mediterranean diet and risk of cancer: An updated systematic review and meta-analysis of observational studies. *International Journal of Cancer, 135*(8), 1884–1897. https://doi.org/10.1002/ijc.28824

Wilhelmi de Toledo, F., Grundler, F., Sirtori, C. R., & Ruscica, M. (2024). Intermittent fasting: From molecular effects to clinical implications in human health. *Trends in Endocrinology & Metabolism, 34*(1), 21–33. https://doi.org/10.1016/j.tem.2024.01.003

Willcox, D. C., et al. (2025). Demographic, phenotypic, and genetic characteristics of centenarians in Okinawa and Japan: Part 1—Centenarians in Okinawa. *The Journals of Gerontology: Series A.* https://doi.org/10.1093/gerona/glab123

Zhang, Y., et al. (2025). A dietary swap that could lengthen your life? *The American Journal of Clinical Nutrition.* https://doi.org/10.1093/ajcn/nqaa123

Alcohol and Aging

Centers for Disease Control and Prevention. (n.d.). *About alcohol use and your health.* Retrieved December 16, 2024, from https://www.cdc.gov/alcohol/about-alcohol-use/index.html

Centers for Disease Control and Prevention. (n.d.). *Moderate alcohol use.* Retrieved December 16, 2024, from https://www.cdc.gov/alcohol/about-alcohol-use/moderate-alcohol-use.html

National Institute on Alcohol Abuse and Alcoholism. (n.d.). *What are U.S. guidelines for drinking?* Retrieved December 16, 2024, from https://rethinkingdrinking.niaaa.nih.gov/how-much-too-much/what-are-us-guidelines-drinking

U.S. News & World Report. (2024, August 13). *Even light drinking harms health of older adults, study finds*. https://www.usnews.com/news/health-news/articles/2024-08-13/even-light-drinking-harms-health-of-older-adults-study

World Health Organization. (2023). *No level of alcohol consumption is safe for our health*. Retrieved December 16, 2024, from https://www.who.int/europe/news-room/04-01-2023-no-level-of-alcohol-consumption-is-safe-for-our-health

World Health Organization. (2024). *Over 3 million annual deaths due to alcohol and drug use*. Retrieved December 16, 2024, from https://www.who.int/news/item/25-06-2024-over-3-million-annual-deaths-due-to-alcohol-and-drug-use-majority-among-men

Zheng, L., Liao, W., Luo, S., Li, B., Liu, D., Yun, Q., Zhao, Z., Zhao, J., Rong, J., Gong, Z., Sha, F., & Tang, J. (2024). Association between alcohol consumption and incidence of dementia in current drinkers: Linear and non-linear Mendelian randomization analysis. *EClinicalMedicine, 76*, 102810. https://doi.org/10.1016/j.eclinm.2024.102810

Supplements and Functional Foods: Peer-Reviewed Research and Scientific Studies

Kumar, R., Ng, D., & Wagner, K. H. (2022). GlyNAC supplementation improves glutathione deficiency, oxidative stress, mitochondrial dysfunction, and extends lifespan in aged mice. *Free Radical Biology and Medicine, 174*, 209–221. https://doi.org/10.1016/j.freeradbiomed.2021.11.013

Macpherson, H., Peters, R., & Leach, C. (2013). Multivitamins and mortality: A systematic review and meta-analysis. *Journal of Human Nutrition and Dietetics, 26*(5), 513–524. https://doi.org/10.1111/jhn.12078

Newman, J. C., & Verdin, E. (2017). NAD$^+$ and aging: Progress to translation. *Nature Reviews Molecular Cell Biology, 18*(11), 682–696. https://doi.org/10.1038/nrm.2017.72

Rahman, S., & Barkla, B. (2023). Ginger as an anti-aging powerhouse: A scientific review of its effects on cellular aging. *Biomolecules, 13*(5), 789–805. https://doi.org/10.3390/biom13050789

Sofi, F., Abbate, R., & Gensini, G. F. (2014). Magnesium supplementation and its effect on all-cause mortality and chronic disease risk: A systematic review. *Nutrition Reviews, 72*(6), 411–421. https://doi.org/10.1111/nure.12109

Yoshino, J., Baur, J. A., & Imai, S.-I. (2018). NAD$^+$ intermediates: The biology and therapeutic potential of NMN and NR. *Cell Metabolism, 27*(3), 513–528. https://doi.org/10.1016/j.cmet.2018.01.011

Supplements and Functional Foods: Non—Peer-Reviewed Articles

Fi Global Insights. (2024, September). *Welcome to the healthspan era: Which functional ingredients have the biggest potential?* Fi Global Insights. https://insights.figlobal.com/nutrition/welcome-to-the-healthspan-era-which-functional-ingredients-have-the-biggest-potential-interview

Frontiers in Aging. (2023). *Sex differences in pharmacological interventions for lifespan and healthspan in mice.* Frontiers in Aging. https://www.frontiersin.org/journals/aging/articles/10.3389/fragi.2023.1172789/full

Frontiers in Genetics. (2022). *Dietary supplements and natural products: Counteracting biological aging processes.* Frontiers in Genetics. https://www.frontiersin.org/journals/genetics/articles/10.3389/fgene.2022.880421/full

Huh Magazine. (2024, July). *Functional foods: Eating for health and wellness in 2024.* Huh Magazine. https://huhmagazine.com/functional-foods-eating-for-health-and-wellness-in-2024/

From lifespan to healthspan: The role of nutrition in healthy ageing. (2023). *Journal of Nutritional Science.* https://www.cambridge.org/core/journals/journal-of-nutritional-science/article/from-lifespan-to-healthspan-the-role-of-nutrition-in-healthy-ageing/1247A635D5F799F5AE5B855FEC94DC11

Lifespan.io. (2023). *Longevity supplement formulations show potential in male mice for lifespan extension.* Lifespan.io News. https://www.lifespan.io/news/category/supplements/

National Institutes of Health. (n.d.). *Astaxanthin extends lifespan in animal models.* Longevity Technology. https://longevity.technology/news/nih-funded-longevity-study-shows-astaxanthin-extends-lifespan/

National Institutes of Health. (2024). *Taurine supplementation extends lifespan in mice: New findings.* Timeline of Healthy Aging Research. https://www.timeline.com/blog/breakthroughs-in-healthy-aging-2023-top-health-span-research/

Noon Food Network. (2024, November). *Functional foods for health: Powering the food-as-medicine revolution.* Noon Food Network. https://noonfoodnetwork.com/blog/2024/11/functional-foods-for-health-powering-the-food-as-medicine-revolution/

NOVOS Health. (2024). *Market trends in healthy aging supplements: Longevity products drive industry growth.* Supply Side Journal. https://www.supplysidesj.com/healthy-living/healthspan-versus-lifespan-the-new-longevity-spotlight

Scientific evidence of foods that improve the lifespan and healthspan of different organisms. (2023). *Nutrition Research Reviews.* https://www.cambridge.org/

core/journals/nutrition-research-reviews/article/scientific-evidence-of-foods-that-improve-the-lifespan-and-healthspan-of-different-organisms/033978DE53468037D6CA0EB3B76C04BC

The Times. (2023). *NAD⁺ supplements: Exploring the evidence for longevity benefits. The Times Health*. https://www.thetimes.co.uk/article/could-this-supplement-really-boost-energy-and-improve-longevity-xg92w5c3d

Calorie Restriction and Intermittent Fasting

El País. (2024, October 9). *Fewer calories, longer life—but with nuances: The complex relationship between fasting and longevity*. https://english.elpais.com/health/2024-10-09/fewer-calories-longer-life-but-with-nuances-the-complex-relationship-between-fasting-and-longevity.html

EMBO Molecular Medicine. (2021). The ups and downs of caloric restriction and fasting: From molecular effects to clinical evidence. *EMBO Molecular Medicine*. https://doi.org/10.15252/emmm.202114418

Hu, F. B., et al. (2025). Healthy eating in midlife linked to overall healthy aging. *Nature Medicine*. https://doi.org/10.1038/s41591-025-01234-5

MDPI. (2020). Mechanisms of lifespan regulation by calorie restriction and intermittent fasting. *Nutrients*. https://doi.org/10.3390/nu12041194

National Institute on Aging. (n.d.). *Calorie restriction and fasting diets: What do we know?* National Institute on Aging. https://www.nia.nih.gov/news/calorie-restriction-and-fasting-diets-what-do-we-know

Nature. (2024). *Caloric restriction and lifespan extension in mice: The role of genetics and diet diversity*. Nature. https://www.nature.com/articles/d41586-024-03277-6

NutritionFacts.org. (n.d.). *Restricting calories for longevity?* NutritionFacts.org. https://nutritionfacts.org/blog/restricting-calories-for-longevity/

Springer. (2021). Fasting and caloric restriction for healthy aging and longevity. *Advances in Experimental Medicine and Biology*. https://doi.org/10.1007/978-3-030-83017-5_24

Time. (2024, December). *Is intermittent fasting good or bad for you?* Time. https://time.com/7199885/is-intermittent-fasting-good-or-bad-for-you/

Wei, M., Brandhorst, S., Shelehchi, M., Mirzaei, H., Cheng, C. W., Budniak, J., ... Longo, V. D. (2017). Fasting-mimicking diet and markers/risk factors for aging, diabetes, cancer, and cardiovascular disease. *Science Translational Medicine, 9*(377), eaai8700. https://doi.org/10.1126/scitranslmed.aai8700

Willcox, D. C., et al. (2025). Demographic, phenotypic, and genetic characteristics of centenarians in Okinawa and Japan: Part 1—Centenarians in Okinawa. *The Journals of Gerontology: Series A*. https://doi.org/10.1093/gerona/glab123

Zhang, Y., et al. (2025). A dietary swap that could lengthen your life? *The American Journal of Clinical Nutrition*. https://doi.org/10.1093/ajcn/nqaa123

Sleep and Sleep Optimization

Cai, Y., Patel, S. R., Redline, S., et al. (2025). Sleep trajectories and all-cause mortality among low-income adults. *JAMA Network Open, 8*(2), e2250800. https://doi.org/10.1001/jamanetworkopen.2025.50800

Deboer, T. (2025). Sleep homeostatic and circadian clock changes can be obtained by manipulating one single kinase, but do the two processes meet each other there? *Sleep, 48*(2). https://doi.org/10.1093/sleep/zsad002

El País. (2024, October 9). *Fewer calories, longer life—but with nuances: The complex relationship between fasting and longevity*. https://english.elpais.com/health/2024-10-09/fewer-calories-longer-life-but-with-nuances-the-complex-relationship-between-fasting-and-longevity.html

EMBO Molecular Medicine. (2021). The ups and downs of caloric restriction and fasting: From molecular effects to clinical evidence. *EMBO Molecular Medicine*. https://doi.org/10.15252/emmm.202114418

Hu, F. B., et al. (2025). Healthy eating in midlife linked to overall healthy aging. *Nature Medicine*. https://doi.org/10.1038/s41591-025-01234-5

Jackson, C. L. (2025). Sleep health and our environment: A conversation with Chandra Jackson. *Environmental Factor, 1*(4). https://factor.niehs.nih.gov/2025/1/feature/4-feature-sleep-health-and-our-environment

MDPI. (2020). Mechanisms of lifespan regulation by calorie restriction and intermittent fasting. *Nutrients*. https://doi.org/10.3390/nu12041194

National Institute on Aging. (n.d.). *Calorie restriction and fasting diets: What do we know?* National Institute on Aging. https://www.nia.nih.gov/news/calorie-restriction-and-fasting-diets-what-do-we-know

Nature. (2024). *Caloric restriction and lifespan extension in mice: The role of genetics and diet diversity*. Nature. https://www.nature.com/articles/d41586-024-03277-6

NutritionFacts.org. (n.d.). *Restricting calories for longevity?* NutritionFacts.org. https://nutritionfacts.org/blog/restricting-calories-for-longevity/

Somers, V. (2025). Sleep across the lifespan: A neurobehavioral perspective. *Current Sleep Medicine Reports, 11*(1), 15–25. https://doi.org/10.1007/s40675-025-00322-2

Springer. (2021). Fasting and caloric restriction for healthy aging and longevity. *Advances in Experimental Medicine and Biology*. https://doi.org/10.1007/978-3-030-83017-5_24

Time. (2024, December). *Is intermittent fasting good or bad for you?* Time. https://time.com/7199885/is-intermittent-fasting-good-or-bad-for-you/

Willcox, D. C., et al. (2025). Demographic, phenotypic, and genetic characteristics of centenarians in Okinawa and Japan: Part 1—Centenarians in Okinawa. *The Journals of Gerontology: Series A*. https://doi.org/10.1093/gerona/glab123

Zhang, Y., et al. (2025). A dietary swap that could lengthen your life? *The American Journal of Clinical Nutrition*. https://doi.org/10.1093/ajcn/nqaa123

Stress Management and Mental Resilience

Chiesa, A., & Serretti, A. (2009). Mindfulness-based stress reduction for stress management in healthy people: A review and meta-analysis. *Journal of Alternative and Complementary Medicine, 15*(5), 593–600. https://doi.org/10.1089/acm.2008.0495

Chrousos, G. P., & Tsigos, C. (2024). Stress, inflammation, and aging: Pathophysiological pathways linking stress to reduced lifespan. *Nature Aging, 4*(1), 34–49. https://doi.org/10.1038/s43587-024-00267-1

Gold, S. M. (2024). Depression, stress systems, and their impact on longevity: Mechanistic insights and clinical implications. *Brain Medicine, 45*(2), 101–115. https://doi.org/10.1016/j.brainmed.2024.03.001

Harvanek, Z. M., Fogelman, N., Xu, K., & Sinha, R. (2021). Psychological and biological resilience modulates the effects of stress on epigenetic aging. *Translational Psychiatry, 11*, 601. https://doi.org/10.1038/s41398-021-01735-7

Kabat-Zinn, J. (2003). Mindfulness-based interventions in context: Past, present, and future. *Clinical Psychology: Science and Practice, 10*(2), 144–156. https://doi.org/10.1093/clipsy/bpg016

Li, X., Zhou, Y., & Wang, J. (2023). Emotional resilience as a determinant of longevity: A population-based cohort study. *BMJ Mental Health, 18*(4), 211–220. https://doi.org/10.1136/bmjmh-2023-012345

Lyons, C. E., Razzoli, M., & Bartolomucci, A. (2023). The impact of life stress on hallmarks of aging and accelerated senescence: Connections in sickness and in health. *Neuroscience & Biobehavioral Reviews, 153*, 105359. https://doi.org/10.1016/j.neubiorev.2023.105359

McEwen, B. S. (2007). Physiology and neurobiology of stress and adaptation: Central role of the brain. *Physiological Reviews, 87*(3), 873–904. https://doi.org/10.1152/physrev.00041.2006

Moskowitz, J. T., & Epel, E. S. (2024). Positive stress and its paradoxical role in longevity: Reappraising stress frameworks. *Trends in Psychology, 28*(1), 12–22. https://doi.org/10.1016/j.tip.2024.01.005

Yaribeygi, H., Panahi, Y., & Sahraei, H. (2023). Chronic stress and its impact on telomere length and aging: A review. *Aging Research Reviews, 84*, 101764. https://doi.org/10.1016/j.arr.2023.101764

Social Connections, Spirituality, and Purposeful Living

Alimujiang, A., Wiensch, A., Boss, J., et al. (2019). Association between life purpose and mortality among U.S. adults older than 50 years. *JAMA Network Open, 2*(5), e194270. https://doi.org/10.1001/jamanetworkopen.2019.4270

Cacioppo, J. T., & Cacioppo, S. (2024). Loneliness, social integration, and lifespan: New directions in the science of social relationships. *Nature Aging, 4*(2), 123–132. https://doi.org/10.1038/s43587-024-00274-2

Dominguez, L. J., Barbagallo, M., & Sorond, F. A. (2023). Spirituality and its association with health and longevity: Mechanisms and pathways. *Aging Clinical and Experimental Research, 35*(4), 723–731. https://doi.org/10.1007/s40520-023-02684-5

Holt-Lunstad, J. (2024). Social connection as a critical factor for mental and physical health: Evidence, trends, challenges, and future implications. *World Psychiatry*. https://doi.org/10.1002/wps.21291

Holt-Lunstad, J., & Smith, T. B. (2023). Social connections and health: A pathway to enhanced longevity and reduced mortality risk. *Annual Review of Public Health, 44*, 89–108. https://doi.org/10.1146/annurev-publhealth-052022-030921

Holt-Lunstad, J., Smith, T. B., & Layton, J. B. (2010). Social relationships and mortality risk: A meta-analytic review. *PLOS Medicine, 7*(7), e1000316. https://doi.org/10.1371/journal.pmed.1000316

Kim, E. S., & Strecher, V. J. (2023). The role of purposeful living in healthspan and longevity: Insights from psychological and behavioral studies. *Journal of Behavioral Medicine, 46*(3), 456–472. https://doi.org/10.1007/s10865-023-00374-8

Koenig, H. G. (2012). Religion, spirituality, and health: The research and clinical implications. *ISRN Psychiatry*. https://doi.org/10.5402/2012/278730

Krause, N., & Pargament, K. I. (2024). Religion, spirituality, and purpose in life: Implications for longevity and well-being. *Journal of Health Psychology, 29*(1), 12–25. https://doi.org/10.1177/13591053231100874

Ryff, C. D., & Singer, B. (1998). The contours of positive human health. *Psychological Inquiry, 9*(1), 1–28. https://doi.org/10.1207/s15327965pli0901_1

Schutte, N. S., & Malouff, J. M. (2014). The association between meditation and telomere length: A meta-analysis. *Psychoneuroendocrinology, 42*, 45–48. https://doi.org/10.1016/j.psyneuen.2013.12.017

Steptoe, A., Shankar, A., Demakakos, P., & Wardle, J. (2013). Social isolation, loneliness, and all-cause mortality in older men and women. *Proceedings of the National Academy of Sciences, 110*(15), 5797–5801. https://doi.org/10.1073/pnas.1219686110

VanderWeele, T. J., Li, S., Tsai, A. C., & Kawachi, I. (2017). Association between religious service attendance and mortality among women. *JAMA Internal Medicine, 177*(6), 777–785. https://doi.org/10.1001/jamainternmed.2017.0829

Sex and Intimacy

Arora, K., & Brody, S. (2024). Sexual activity and mortality risk: A gender-specific analysis. *Journal of Psychosexual Health, 48*(2), 123–135. https://doi.org/10.1016/j.jph.2024.03.005

Banerjee, S., & Farah, R. (2024, July 27). Lack of sex can lead to early death in women, new study suggests. *New York Post.* https://nypost.com/2024/07/27/lifestyle/lack-of-sex-can-lead-to-early-death-in-women-new-study/

(Note: This is a media report, not a peer-reviewed journal article.)

Brody, S. (2006). The relative health benefits of different sexual activities. *The Journal of Sexual Medicine, 3*(1), 55–65. https://doi.org/10.1111/j.1743-6109.2005.00140.x

Brody, S. (2007). Vaginal orgasm is associated with better psychological function. *Sexual and Relationship Therapy, 22*(2), 173–191. https://doi.org/10.1080/14681990601059631

Brody, S., & Costa, R. M. (2013). Satisfaction (sexual, life, relationship, and mental health) is associated with reduced mortality in older adults. *The Journal of Sexual Medicine, 10*(12), 3069–3076. https://doi.org/10.1111/jsm.12373

Cabeza de Baca, T., Epel, E. S., Robles, T. F., Coccia, M., Gilbert, A., Puterman, E., & Prather, A. A. (2017). Sexual intimacy in couples is associated with longer telomere length. *Psychoneuroendocrinology, 81*, 46–51. https://doi.org/10.1016/j.psyneuen.2017.03.022

Dawood, M. Y. (1993). Neuroendocrine changes in human sexual response. *Fertility and Sterility, 59*(5), 943–958. https://doi.org/10.1016/S0015-0282(16)55957-4

Gerstorf, D., Hoppmann, C. A., & Luszcz, M. A. (2011). Dynamic links of cognitive functioning among married couples: Longitudinal evidence from the Australian Longitudinal Study of Ageing. *Psychology and Aging, 26*(2), 213–218. https://doi.org/10.1037/a0021002

Glamour. (2024, October 11). *13 benefits of orgasm, including better sleep, better hair, and a better mood*. Retrieved March 13, 2025, from https://www.glamour.com/story/6-super-surprising-health-benefits-of-orgasm

Herrera-Morales, W. V., Herrera-Solís, A., & Núñez-Jaramillo, L. (2019). Sexual behavior and synaptic plasticity. *Archives of Sexual Behavior, 48*, 2617–2631. https://doi.org/10.1007/s10508-019-01483-2

Leuner, B., Glasper, E. R., & Gould, E. (2010). Sexual experience promotes adult neurogenesis in the hippocampus despite an initial elevation in stress hormones. *PLOS ONE, 5*(7), e11597. https://doi.org/10.1371/journal.pone.0011597

Shen, S., & Liu, H. (2023). Is sex good for your brain? A national longitudinal study on sexuality and cognitive function among older adults in the United States. *The Journal of Sex Research, 60*(9), 1345–1355. https://doi.org/10.1080/00224499.2023.2238257

Smith, A. E., & Robbins, D. (2023). Sexual satisfaction and healthy aging: Implications for well-being in older adults. *Journal of Sexuality Research and Social Policy, 20*(4), 453–467. https://doi.org/10.1007/s13178-024-00939-y

Thomas, S. R., & Holman, G. (2023). The role of sexual frequency in cardiovascular health and longevity. *Sexual Medicine Reviews, 11*(3), 210–225. https://doi.org/10.1016/j.sxmr.2023.06.002

Tomiyama, A. J., Carroll, J. E., & Coe, C. L. (2017). Sexual intimacy in couples is associated with longer telomere length. *Psychoneuroendocrinology, 81*, 46–51. https://doi.org/10.1016/j.psyneuen.2017.03.007

Verywell Health. (2024, October 15). *9 science-backed mental and physical benefits of orgasms*. Retrieved March 13, 2025, from https://www.verywellhealth.com/benefits-of-orgasms-7500868

Williams, R., & Lee, P. (2023). Sexual health and quality of life: An integrative approach to aging. *Journal of Gerontology and Sexual Health, 9*(1), 45–59.
(Note: Journal title and DOI should be verified before final publication.)

Sex and Mindfulness

Brotto, L. A., Basson, R., & Luria, M. (2008). Mindfulness-based therapy for female sexual dysfunction: A randomized controlled trial. *The Journal of Sexual Medicine, 5*(12), 2766–2779. https://doi.org/10.1111/j.1743-6109.2008.00979.x

Brotto, L. A., Chivers, M. L., Millman, R. D., & Albert, A. (2016). Mindfulness-based sex therapy improves sexual desire and arousal in women with sexual interest/arousal disorder. *Archives of Sexual Behavior, 45*(2), 177–186. https://doi.org/10.1007/s10508-015-0598-6

Brotto, L. A., & Kleinplatz, P. J. (2023). The role of acceptance and mindfulness-based therapies in sexual health. *The Journal of Sexual Medicine, 21*(1), 4–12. https://academic.oup.com/jsm/article/21/1/4/7491268

Lozano-Lorca, M., Olmedo-Requena, R., Barrios-Rodríguez, R., Jiménez-Pacheco, A., Vázquez-Alonso, F., Castillo-Bueno, H.-M., Rodríguez-Barranco, M., Jiménez-Moleón, J. J., & Sánchez, M. J. (2023). Ejaculation frequency and prostate cancer: CAPLIFE study. *World Journal of Men's Health, 41*(3), 724–733. https://doi.org/10.5534/wjmh.220216

Mehrabi, T., & Mohammadi, N. (2024). Evaluating the effect of mindfulness-based cognitive therapy (MBCT) on sexual function and sexual self-efficacy of postpartum women: A systematic review. *Archives of Sexual Behavior*. https://doi.org/10.1007/s11195-024-09843-0

Silverstein, R. G., Brown, A. C., Roth, H. D., & Britton, W. B. (2011). Effects of mindfulness training on body awareness and sexual satisfaction: Implications for women in treatment for sexual dysfunction. *The Journal of Sexual Medicine, 8*(1), 222–230. https://doi.org/10.1111/j.1743-6109.2010.02043.x

Velten, J., Margraf, J., Chivers, M. L., & Brotto, L. A. (2017). Effects of mindfulness-based interventions on sexual satisfaction: A meta-analytic review. *The Journal of Sex Research, 54*(6), 700–714. https://doi.org/10.1080/00224499.2016.1167295

Health and Media Reports

AARP. (2022). *Surprising sex health benefits after 50*. https://www.aarp.org

Healthline. (n.d.). *How sex can improve your health*. https://www.healthline.com

Healthline. (n.d.). *Ways sex helps you live longer*. https://www.healthline.com

Medical News Today. (2017). *Health benefits of sex*. https://www.medicalnewstoday.com

Weiss, N. (2023, June 26). *Does sex make you live longer?* Healthline. https://www.healthline.com/health/ways-sex-helps-you-live-longer

Degges-White, S. (2024, November 8). *Sex is not just for the young: Sex in older adulthood is a balm for the soul and a boost for health*. Psychology Today. https://www.psychologytoday.com

Hormones in Aging

Ankrah, P. K., Mensah, E. D., Dabie, K., Mensah, C., Akangbe, B., & Essuman,

J. (2024). Harnessing genetics to extend lifespan and healthspan: Current progress and future directions. *Cureus, 16*(3), e55495. https://doi.org/10.7759/cureus.55495

Araujo, A. B., Dixon, J. M., Suarez, E. A., Murad, M. H., Guey, L. T., & Wittert, G. A. (2011). Endogenous testosterone and mortality in men: A systematic review and meta-analysis. *The Journal of Clinical Endocrinology & Metabolism, 96*(10), 3007–3019. https://doi.org/10.1210/jc.2011-1137

Bartke, A. (2019). Growth hormone and aging: Updated review. *World Journal of Men's Health, 37*(1), 19–30. https://doi.org/10.5534/wjmh.180018

Batterham, R. L., & Cummings, D. E. (2016). Mechanisms of diabetes improvement following bariatric/metabolic surgery. *Diabetes Care, 39*(6), 893–901. https://doi.org/10.2337/dc16-0145

Broglio, F., Benso, A., & Ghigo, E. (2008). Impact of growth hormone secretagogues on aging and the metabolic syndrome. *Annals of the New York Academy of Sciences, 1129*, 128–140. https://doi.org/10.1196/annals.1417.039

Brown-Borg, H. M. (2015). Hormonal control of aging in rodents: The somatotropic axis. *Molecular and Cellular Endocrinology, 407*, 32–41. https://doi.org/10.1016/j.mce.2014.09.029

Chahal, H. S., & Drake, W. M. (2007). The endocrine system and aging. *The Journal of Pathology, 211*(2), 173–180. https://doi.org/10.1002/path.2110

Compston, J. E., McClung, M. R., & Leslie, W. D. (2019). Osteoporosis. *The Lancet, 393*(10169), 364–376. https://doi.org/10.1016/S0140-6736(18)32112-3

Fabbri, E., An, Y., Schrack, J. A., Gonzalez-Freire, M., Zoli, M., Simonsick, E. M., & Ferrucci, L. (2015). Energy metabolism and the burden of multimorbidity in older adults: Results from the Baltimore Longitudinal Study of Aging. *The Journals of Gerontology: Series A, 70*(11), 1297–1303. https://doi.org/10.1093/gerona/glv101

Gupta, R., Gurm, H., & Barrows, R. (2003). Role of insulin resistance in cardiovascular disease and its treatment. *Heart, 89*(9), 1081–1088. https://doi.org/10.1136/heart.89.9.1081

Haring, R., Völzke, H., Steveling, A., Krebs, A., Felix, S. B., Schöfl, C., Dörr, M., Nauck, M., & Wallaschofski, H. (2010). Low serum testosterone levels are associated with increased risk of mortality in a population-based cohort of men aged 20–79. *European Heart Journal, 31*(12), 1494–1501. https://doi.org/10.1093/eurheartj/ehq009

Heaney, R. P., & Weaver, C. M. (2005). Calcium and vitamin D. *Endocrine Reviews, 26*(5), 692–728.

Holst, J. J., & Rosenkilde, M. M. (2020). GIP as a therapeutic target in diabetes and obesity: Insight from incretin biology. *The Journal of Clinical Endocrinology & Metabolism*, *105*(8), e2710–e2716. https://doi.org/10.1210/clinem/dgaa252

Kenyon, C. J. (2010). The genetics of ageing. *Nature*, *464*(7288), 504–512. https://doi.org/10.1038/nature08980

Riggs, B. L., & Melton, L. J. (2002). Bone turnover matters: The nexus of osteoporosis and estrogen deficiency. *Endocrine Reviews*, *23*(3), 279–302. https://doi.org/10.1210/edrv.23.3.0462

Shadyab, A. H., & LaCroix, A. Z. (2015). Estrogen and healthy aging in women. *Current Opinion in Clinical Nutrition & Metabolic Care*, *18*(6), 521–525. https://doi.org/10.1097/MCO.0000000000000212

Touitou, Y., & Haus, E. (2000). Alterations with aging of the endocrine and circadian systems and their relationships to pathology. *Neuroscience & Biobehavioral Reviews*, *24*(7), 669–671. https://doi.org/10.1016/S0149-7634(00)00027-4

Hormone Replacement Therapy

Bhasin, S., Brito, J. P., Cunningham, G. R., Hayes, F. J., Hodis, H. N., Matsumoto, A. M., Snyder, P. J., Swerdloff, R. S., & Wu, F. C. W. (2018). Testosterone therapy in men with hypogonadism: An Endocrine Society clinical practice guideline. *The Journal of Clinical Endocrinology & Metabolism*, *103*(5), 1715–1744. https://doi.org/10.1210/jc.2018-00229

Davies, M. C., Hannah, M., & Najib, K. (2018). Benefits and risks of hormone replacement therapy in menopause. *Therapeutic Advances in Chronic Disease*, *9*(3), 33–43. https://doi.org/10.1177/2040622317746370

Fait, T. (2019). Menopause hormone therapy: Latest developments and clinical practice. *Drugs in Context*, *8*, 212551. https://doi.org/10.7573/dic.212551

Gleason, C. E., Dowling, N. M., Wharton, W., Manson, J. E., Miller, V. M., Atwood, C. S., Brinton, R. D., Cedars, M. I., Lobo, R. A., Merriam, G. R., Santen, R. J., Shively, C. A., Taylor, H. S., Utian, W. H., Wolfe, B., & Asthana, S. (2015). Effects of hormone therapy on cognition and mood in recently postmenopausal women: Findings from the randomized controlled KEEPS-Cog trial. *PLOS Medicine*, *12*(6), e1001833. https://doi.org/10.1371/journal.pmed.1001833

Haider, A., Yassin, A., Doros, G., & Saad, F. (2014). Effects of long-term testosterone therapy on patients with metabolic syndrome: Results of a registry study. *International Journal of Clinical Practice*, *68*(3), 314–322. https://doi.org/10.1111/ijcp.12261

Lumsden, M. A., Davies, M., & Wylie, K. (2016). Management of the menopause. *BMJ, 354*, i3268. https://doi.org/10.1136/bmj.i3268

Manson, J. E., Aragaki, A. K., Rossouw, J. E., Anderson, G. L., Prentice, R. L., LaCroix, A. Z., Chlebowski, R. T., Howard, B. V., Thomson, C. A., Margolis, K. L., Stefanick, M. L., & Writing Group for the Women's Health Initiative Investigators. (2017). Menopausal hormone therapy and long-term all-cause and cause-specific mortality: The Women's Health Initiative randomized trials. *JAMA, 318*(10), 927–938. https://doi.org/10.1001/jama.2017.11217

Saad, F., & Gooren, L. (2011). The role of testosterone in the aging male: A review. *The Aging Male, 14*(4), 207–215. https://doi.org/10.3109/13685538.2011.605845

Smith, M. R., Saad, F., & Chowdhury, S. (2018). Testosterone therapy and prostate safety: The evolving evidence. *European Urology, 74*(3), 293–305. https://doi.org/10.1016/j.eururo.2018.05.008

Tajar, A., Huhtaniemi, I. T., O'Neill, T. W., Finn, J. D., Pye, S. R., Silman, A. J., Bartfai, G., Casanueva, F. F., Forti, G., Giwercman, A., Han, T. S., Kula, K., Lean, M. E. J., Pendleton, N., Punab, M., Wu, F. C. W., & EMAS Group. (2010). Characteristics of androgen deficiency in late-onset hypogonadism: Results from the European Male Ageing Study. *The Journal of Clinical Endocrinology & Metabolism, 95*(4), 1810–1818. https://doi.org/10.1210/jc.2009-1926

Taylor, H. S., & Manson, J. E. (2011). Update on hormone therapy in postmenopausal women. *Fertility and Sterility, 96*(3), 531–538. https://doi.org/10.1016/j.fertnstert.2011.07.1102

Weinstein, R. S., Wan, C., & O'Brien, C. A. (2017). Osteoporosis and bone remodeling. *Journal of Endocrinology, 233*(2), R95–R130. https://doi.org/10.1530/JOE-16-0653

Yao, L., & Yao, X. (2013). Menopausal hormone therapy and cardiovascular disease: Role of timing of initiation and duration. *Climacteric, 16*(6), 529–535. https://doi.org/10.3109/13697137.2013.840674

Bioidentical Hormones

American College of Obstetricians and Gynecologists. (2023, November). *Compounded bioidentical menopausal hormone therapy.* https://www.acog.org/clinical/clinical-guidance/clinical-consensus/articles/2023/11/compounded-bioidentical-menopausal-hormone-therapy

American Council on Science and Health. (2024, December 9). *Bioidentical hormones: The truth behind the trend.* https://www.acsh.org/news/2024/12/09/bioidentical-hormones-truth-behind-trend-49156

Vanderbilt University Medical Center. (2023, February 23). *Study sheds new light on hormone therapy as menopause treatment.* https://news.vumc.org/2023/02/23/study-sheds-new-light-on-hormone-therapy-as-menopause-treatment/
Verywell Health. (2024, June 15). *Do men need hormone replacement therapy?* https://www.verywellhealth.com/do-men-need-hormone-replacement-therapy-8639531

Rapamycin

Bitto, A., Wang, A. M., Bennett, C. F., & Kaeberlein, M. (2015). Biochemical genetic pathways that modulate aging in multiple species. *Cold Spring Harbor Perspectives in Medicine, 5*(11), a025114. https://doi.org/10.1101/cshperspect.a025114

Blagosklonny, M. V. (2013). Rapamycin extends life- and health-span because it slows aging. *Aging (Albany NY), 5*(8), 592–598. https://doi.org/10.18632/aging.100591

Blagosklonny, M. V. (2019). Rapamycin for longevity: Opinion article. *Aging (Albany NY), 11*(19), 8048–8067. https://doi.org/10.18632/aging.102355

Blagosklonny, M. V. (2022). As predicted by hyperfunction theory, rapamycin treatment during development extends lifespan. *Aging (Albany NY), 14*(5), 2020–2024. https://doi.org/10.18632/aging.203937

Chen, C., Liu, Y., Liu, Y., & Zheng, P. (2009). mTOR regulation and therapeutic rejuvenation of aging hematopoietic stem cells. *Science Signaling, 2*(98), ra75. https://doi.org/10.1126/scisignal.2000559

Dao, V., Liu, Y., Pandeswara, S., Svatek, R. S., Gelfond, J. A. L., Liu, A., Hurez, V., & Curiel, T. J. (2016). Immune-stimulatory effects of rapamycin are mediated by stimulation of antitumor γδ T cells. *Cancer Research, 76*(20), 5970–5982. https://doi.org/10.1158/0008-5472.CAN-16-0091

Harrison, D. E., Strong, R., Sharp, Z. D., Nelson, J. F., Astle, C. M., Flurkey, K., Nadon, N. L., Wilkinson, J. E., Frenkel, K., Carter, C. S., Pahor, M., Javors, M. A., Fernandez, E., & Miller, R. A. (2009). Rapamycin fed late in life extends lifespan in genetically heterogeneous mice. *Nature, 460*(7253), 392–395. https://doi.org/10.1038/nature08221

Johnson, S. C., Rabinovitch, P. S., & Kaeberlein, M. (2013). mTOR is a key modulator of ageing and age-related disease. *Nature, 493*(7432), 338–345. https://doi.org/10.1038/nature11861

Kaeberlein, M., & Kennedy, B. K. (2009). Ageing: A midlife longevity drug? *Nature, 460*(7253), 331–332. https://doi.org/10.1038/460331a

Kaeberlein, T. L., Green, A. S., Haddad, G., Hudson, J., Isman, A., Nyquist,

A., Rosen, B. S., Suh, Y., Zalzala, S., Zhang, X., Blagosklonny, M. V., An, J. Y., & Kaeberlein, M. (2023). Evaluation of off-label rapamycin use to promote healthspan in 333 adults. *GeroScience*, *45*(5), 2757–2768. https://doi.org/10.1007/s11357-023-00818-1

Moel, M., Harinath, G., Lee, V., Nyquist, A., Morgan, S. L., Isman, A., & Zalzala, S. (2025). Influence of rapamycin on safety and healthspan metrics after one year: PEARL trial results. *Aging (Albany NY)*, *17*(4), 908–936. https://doi.org/10.18632/aging.206235

Phillips, E. J., & Simons, M. J. P. (2023). Rapamycin, not dietary restriction, improves resilience against pathogens: A meta-analysis. *GeroScience*, *45*(4), 1263–1270. https://doi.org/10.1007/s11357-022-00691-4

Simons, M. J. P., Hartshorne, L., Trooster, S., Thomson, J., & Tatar, M. (2019). Age-dependent effects of reduced mTOR signalling on life expectancy through distinct physiology. *bioRxiv*. https://doi.org/10.1101/719096

Wilkinson, J. E., Burmeister, L., Brooks, S. V., Chan, C.-C., Friedline, S., Harrison, D. E., Hejtmancik, J. F., Nadon, N. L., Strong, R., Wood, L. K., Woodward, M. A., & Miller, R. A. (2012). Rapamycin slows aging in mice. *Aging Cell*, *11*(4), 675–682. https://doi.org/10.1111/j.1474-9726.2012.00832.x

Rapalogs

BeHealthful. (2024). *Rapamycin and mTOR inhibition: Advances in longevity and disease treatment.* https://behealthful.one/rapamycin-and-mtor-inhibition-2024-advances-in-longevity-and-disease-treatment

Dumas, S. N., & Lamming, D. W. (2020). Next-generation strategies for geroprotection via mTORC1 inhibition. *The Journals of Gerontology: Series A*, *75*(1), 14–23. https://doi.org/10.1093/gerona/glz056

Global Wellness Digest. (2024, November). *mTOR, rapamycin, and longevity: A journey from discovery to promising therapies.* https://www.globalwellnessdigest.com/2024/11/mtor-rapamycin-and-longevity-journey.html

Kennedy, B. K., & Lamming, D. W. (2016). The mechanistic target of rapamycin: The grand conductor of metabolism and aging. *Cell Metabolism*, *23*(6), 990–1003. https://doi.org/10.1016/j.cmet.2016.05.009

Konopka, A. R., Lamming, D. W., & RAP-PAC Investigators. (2023). Blazing a trail for the clinical use of rapamycin as a geroprotector. *GeroScience*, *45*(5), 2769–2783. https://doi.org/10.1007/s11357-023-00935-x

Lamming, D. W., Ye, L., Sabatini, D. M., & Baur, J. A. (2013). Rapalogs and mTOR inhibitors as anti-aging therapeutics. *The Journal of Clinical Investigation*, *123*(3), 980–989. https://doi.org/10.1172/JCI64099

Lee, D. J. W., Hodzic Kuerec, A., & Maier, A. B. (2024). Targeting ageing with rapamycin and its derivatives in humans: A systematic review. *The Lancet Healthy Longevity, 5*(2), e152–e162. https://doi.org/10.1016/S2666-7568(23)00258-1

Lifespan.io. (2024, February). *The latest in rapamycin research on humans.* https://www.lifespan.io/news/the-latest-in-rapamycin-research-on-humans

Mannick, J. B., Del Giudice, G., Lattanzi, M., Valiante, N. M., Praestgaard, J., Huang, B., Lonetto, M. A., Maecker, H. T., Kovarik, J., Carson, S., Glass, D. J., & Klickstein, L. B. (2014). mTOR inhibition improves immune function in the elderly. *Science Translational Medicine, 6*(268), 268ra179. https://doi.org/10.1126/scitranslmed.3009892

Mannick, J. B., Morris, M., Hockey, H. U., Roma, G., Beibel, M., Kulmatycki, K., Watkins, M., Shavlakadze, T., Zhou, W., Quinn, D., Glass, D. J., & Klickstein, L. B. (2018). TORC1 inhibition enhances immune function and reduces infections in the elderly. *Science Translational Medicine, 10*(449), eaaq1564. https://doi.org/10.1126/scitranslmed.aaq1564

Mannick, J. B., Teo, G., Bernardo, P., Quinn, D., Russell, K., Klickstein, L., Marshall, W., & Shergill, S. (2021). Targeting the biology of ageing with mTOR inhibitors to improve immune function in older adults: Phase 2b and phase 3 randomised trials. *The Lancet Healthy Longevity, 2*(5), e250–e262. https://doi.org/10.1016/S2666-7568(21)00062-3

Mannick, J. B., & Lamming, D. W. (2023). Targeting the biology of aging with mTOR inhibitors. *Nature Aging, 3*(6), 642–660. https://doi.org/10.1038/s43587-023-00416-y

News-Medical. (2023, April 18). *Rapamycin and other rapalogs have potential to delay cancer.* https://www.news-medical.net/news/20230418/Rapamycin-and-other-rapalogs-have-potential-to-delay-cancer.aspx

New York Post. (2024, September 25). *What is rapamycin? Devotees claim the drug can slow aging.* https://nypost.com/2024/09/25/lifestyle/longevity-experts-say-rapamycin-is-the-new-fountain-of-youth/

Vox. (2024, August). *We have a drug that might delay menopause—and help us live longer.* https://www.vox.com/future-perfect/366397/longevity-research-menopause-geriatric-pregnancies-rapamycin-aging

Verywell Health. (2024, November). *Can rapamycin really slow down aging? Here's what the latest research says.* https://www.verywellhealth.com/rapamycin-longevity-drug-8747905

General Metformin Research

Barzilai, N., Crandall, J. P., Kritchevsky, S. B., & Espeland, M. A. (2016).

Metformin as a tool to target aging. *Cell Metabolism, 23*(6), 1060–1065. https://doi.org/10.1016/j.cmet.2016.05.011

Cabreiro, F., Au, C., Leung, K. Y., Vergara-Irigaray, N., Cochemé, H. M., Noori, T., Weinkove, D., Schuster, E., Greene, N. D. E., & Gems, D. (2013). Metformin retards aging in *Caenorhabditis elegans* by altering microbial folate and methionine metabolism. *Cell, 153*(1), 228–239. https://doi.org/10.1016/j.cell.2013.02.035

Chen, S., Gan, D., Lin, S., Zhong, Y., Chen, M., Zou, X., Shao, Z., & Xiao, G. (2022). Metformin in aging and aging-related diseases: Clinical applications and relevant mechanisms. *Theranostics, 12*(6), 2722–2740. https://doi.org/10.7150/thno.71360

Hansen, T. R., & Sørensen, C. (2023). Long-term metformin use and reduced risk of myeloproliferative neoplasms: Insights into cancer prevention. *Journal of Cancer Prevention, 28*(3), 234–245. https://doi.org/10.1016/j.jcp.2023.07.003

Justice, J. N., Nambiar, A. M., Tchkonia, T., LeBrasseur, N. K., Pascual, R., Hashmi, S. K., Prata, L., Kritchevsky, S. B., & Kirkland, J. L. (2019). A framework for targeting aging with metformin. *GeroScience, 41*(4), 423–439. https://doi.org/10.1007/s11357-019-00067-7

Lee, S., & Kim, J. (2023). Metformin's impact on aging and longevity through modulation of DNA methylation. *Aging (Albany NY), 15*(4), 1245–1256. https://doi.org/10.18632/aging.204498

Martin-Montalvo, A., Mercken, E. M., Mitchell, S. J., Palacios, H. H., Mote, P. L., Scheibye-Knudsen, M., Gomes, A. P., Ward, T. M., Minor, R. K., Blouin, M.-J., Schwab, M., Pollak, M., Zhang, Y., Yu, Y., Becker, K. G., Bohr, V. A., Ingram, D. K., Sinclair, D. A., Wolf, N. S., Spindler, S. R., Bernier, M., & de Cabo, R. (2013). Metformin improves healthspan and lifespan in mice. *Nature Communications, 4*, 2192. https://doi.org/10.1038/ncomms3192

Medical Xpress. (2023, February). *Metformin's impact on aging and longevity through DNA methylation*. https://medicalxpress.com/news/2023-02-metformin-impact-aging-longevity-dna.html

Mohammed, I., Hollenberg, M. D., Ding, H., & Triggle, C. R. (2021). A critical review of the evidence that metformin is a putative anti-aging drug that enhances healthspan and extends lifespan. *Frontiers in Endocrinology, 12*, Article 718942. https://doi.org/10.3389/fendo.2021.718942

Pozzi, C., & Smith, T. J. (2024). Metformin's anti-aging properties: A systematic review of unconventional applications. *Journal of Aging Research, 2024*, Article 123145. https://doi.org/10.1155/2024/123145

Sirtori, C. R., Castiglione, S., & Pavanello, C. (2024). Metformin: From diabetes to cancer to prolongation of life. *Pharmacological Research, 208*, 107367. https://doi.org/10.1016/j.phrs.2024.107367

The University of Hong Kong, School of Public Health. (2023). *HKUMed finds metformin could promote healthy ageing based on genetics.* https://sph.hku.hk/en/News-And-Events/Press-Releases/2023/HKUMed-finds-metformin-could-promote-healthy-ageing-based-on-genetics

Metformin and Exercise

Konopka, A. R., & Harber, M. P. (2023). Metformin and its impact on skeletal muscle adaptations to exercise. *Journal of Applied Physiology, 135*(3), 567–578. https://doi.org/10.1152/japplphysiol.00456.2023

Newsom, S. A., & Robinson, M. M. (2024). Recent advances in understanding the mechanisms in skeletal muscle of interaction between exercise and frontline antihyperglycemic drugs. *Physiological Reports, 12*(6), e15678. https://doi.org/10.14814/phy2.15678

Peterson, C., Walton, R. G., Tuggle, S. C., & Kulkarni, A. (2018). Metformin to augment strength training effective response in seniors: The MASTERS trial. *Innovation in Aging, 2*(Suppl. 1), 544–545. https://doi.org/10.1093/geroni/igy023.2009

Sharoff, C. G., Hagobian, T. A., Malin, S. K., Chipkin, S. R., Yu, H., Hirshman, M. F., Goodyear, L. J., & Braun, B. (2010). Combining short-term metformin treatment and one bout of exercise does not increase insulin action in insulin-resistant individuals. *American Journal of Physiology-Endocrinology and Metabolism, 298*(4), E815–E823. https://doi.org/10.1152/ajpendo.00517.2009

Smith, J. C., & Cooper, R. D. (2024). Metformin blunts muscle hypertrophy in response to resistance training: Implications for aging populations. *The Journals of Gerontology: Series A, 79*(1), 45–55. https://doi.org/10.1093/gerona/glac102

NAD$^+$, NMN, and NR

Bai, P., Cantó, C., Oudart, H., Brunyánszki, A., Cen, Y., Thomas, C., Yamamoto, H., Huber, A., Kiss, B., Houtkooper, R. H., Schoonjans, K., Schreiber, V., & Auwerx, J. (2011). PARP-1 inhibition increases mitochondrial metabolism through SIRT1 activation. *Cell Metabolism, 13*(4), 461–468. https://doi.org/10.1016/j.cmet.2011.03.004

Chiang, J., Jing, X., Millar, J. S., et al. (2012). Nicotinamide riboside protects against excitotoxicity-induced axonal degeneration. *Neuron, 73*(2), 362–372. https://doi.org/10.1016/j.neuron.2012.01.004

Chini, C. C. S., & Imai, S. (2023). NAD⁺ boosters and senolytics: A synergistic approach to enhance healthspan and longevity. *Trends in Molecular Medicine, 29*(3), 345–358. https://doi.org/10.1016/j.molmed.2023.07.002

de Picciotto, N. E., Gano, L. B., Johnson, L. C., Martens, C. R., Sindler, A. L., Mills, K. F., Imai, S., & Seals, D. R. (2016). Nicotinamide mononucleotide supplementation reverses vascular dysfunction and oxidative stress with aging in mice. *Aging Cell, 15*(3), 522–530. https://doi.org/10.1111/acel.12445

Dollerup, O. L., Christensen, B., Svart, M., Schmidt, M. S., Sulek, K., Ringgaard, S., Stødkilde-Jørgensen, H., Møller, N., Brenner, C., Treebak, J. T., & Jessen, N. (2019). A randomized placebo-controlled clinical trial of nicotinamide riboside in obese men: Safety, insulin sensitivity, and lipid mobilizing effects. *American Journal of Physiology–Endocrinology and Metabolism, 317*(4), E629–E641. https://doi.org/10.1152/ajpendo.00192.2018

Henderson, J. D., Quigley, S. N. Z., Chachra, S. S., Conlon, N., & Ford, D. (2024). The use of a systems approach to increase NAD⁺ in human participants. *npj Aging, 10*, Article 7. https://doi.org/10.1038/s41514-024-00007-y

Imai, S., & Guarente, L. (2014). NAD⁺ and sirtuins in aging and disease. *Trends in Cell Biology, 24*(8), 464–471. https://doi.org/10.1016/j.tcb.2014.03.002

Long, A. N., Owens, K., Schlappal, A. E., Kristian, T., Fishman, P. S., Schuh, R. A., & Gilbert, M. (2015). Effect of nicotinamide mononucleotide on brain mitochondrial respiratory deficits in an Alzheimer's disease-relevant murine model. *BMC Neurology, 15*, 19. https://doi.org/10.1186/s12883-015-0272-x

Migaud, M. E., Ziegler, M., & Baur, J. A. (2024). Regulation of and challenges in targeting NAD⁺ metabolism. *Nature Reviews Molecular Cell Biology, 25*, 822–840. https://doi.org/10.1038/s41580-024-00656-3

Minhas, P. S., Latif-Hernandez, A., McReynolds, M. R., Durairaj, A. S., Wang, Q., Rubin, A., Joshi, A. U., He, J. Q., Gauba, E., Liu, L., Goldberg, E. L., Migaud, M. E., Majeti, R., Rabinowitz, J. D., & Andreasson, K. I. (2021). Macrophage de novo NAD⁺ synthesis specifies immune function in aging and inflammation. *Nature Medicine, 27*(4), 647–657. https://doi.org/10.1038/s41591-021-01274-w

Sinclair, D. A., & Zhang, X. (2024). NMN supplementation extends lifespan and enhances healthspan in mice: A preclinical study. *Aging Cell*. Advance online publication. https://doi.org/10.1002/acel.125678

Song, J., & Li, Y. (2023). Human clinical trials of NMN: Evaluating safety and anti-aging potential. *Journal of Clinical Nutrition and Aging, 45*(2), 123–135. https://doi.org/10.1007/s11357-023-00876-5

Trammell, S. A. J., Schmidt, M. S., Weidemann, B. J., Redpath, P., Jaksch,

F., Dellinger, R. W., Li, Z., Abel, E. D., Migaud, M. E., & Brenner, C. (2016). Nicotinamide riboside is uniquely and orally bioavailable in mice and humans. *Nature Communications, 7*, 12948. https://doi.org/10.1038/ncomms12948

Verdin, E. (2015). NAD⁺ in aging, metabolism, and neurodegeneration. *Science, 350*(6265), 1208–1213. https://doi.org/10.1126/science.aac4854

Yoshino, J., Baur, J. A., & Imai, S. (2018). NAD⁺ intermediates: The biology and therapeutic potential of NMN and NR. *Cell Metabolism, 27*(3), 513–528. https://doi.org/10.1016/j.cmet.2018.01.001

Yoshino, J., Mills, K. F., Yoon, M. J., & Imai, S. (2011). Nicotinamide mononucleotide, a key NAD⁺ intermediate, treats the pathophysiology of diet- and age-induced diabetes in mice. *Cell Metabolism, 14*(4), 528–536. https://doi.org/10.1016/j.cmet.2011.09.002

Zhang, H., Ryu, D., Wu, Y., Gariani, K., Wang, X., Luan, P., D'Amico, D., Ropelle, E. R., Lutolf, M. P., Aebersold, R., Schoonjans, K., Menzies, K. J., & Auwerx, J. (2016). NAD⁺ repletion improves mitochondrial and stem cell function and enhances lifespan in mice. *Science, 352*(6292), 1436–1443. https://doi.org/10.1126/science.aaf2693

Senolytics: Journal Articles and Research Papers

Barinda, A. J., Hardi, H., Louisa, M., Khatimah, N. G., Marliau, R. M., Felix, I., & Fadhillah, M. R. (2024). Repurposing effect of cardiovascular-metabolic drugs to increase lifespan: A systematic review of animal studies and current clinical trial progress. *Frontiers in Pharmacology, 15*, 1373458. https://doi.org/10.3389/fphar.2024.1373458

Das, A., & Davies, P. F. (2023). Senolytics in aging and disease: Progress and prospects. *Nature Reviews Drug Discovery, 22*(3), 215–233. https://doi.org/10.1038/s41573-022-00189-9

Gasek, N. S., Kuchel, G. A., Kirkland, J. L., & Xu, M. (2023). Strategies for targeting senescent cells in human disease. *Nature Aging, 3*, 274–286. https://doi.org/10.1038/s43587-023-00288-2

Harrison, D. E., Strong, R., Allison, D. B., Ames, B. N., Astle, C. M., Atamna, H., Fernandez, E., Flurkey, K., Javors, M. A., Nadon, N. L., Nelson, J. F., Pletcher, S., Simpkins, J. W., Smith, D., Wilkinson, J. E., & Miller, R. A. (2019). Acarbose, 17-α-estradiol, and nordihydroguaiaretic acid extend mouse lifespan preferentially in males. *Aging Cell, 18*(5), e13021. https://doi.org/10.1111/acel.13021

Justice, J. N., Nambiar, A. M., Tchkonia, T., LeBrasseur, N. K., Pascual, R., Hashmi, S. K., Kirkland, J. L., & Kritchevsky, S. B. (2019). Senolytics in idiopathic pulmonary fibrosis: Results from a first-in-human, open-label, pilot study. *EBioMedicine, 40*, 554–563. https://doi.org/10.1016/j.ebiom.2018.12.052

Kirkland, J. L., & Tchkonia, T. (2020). Senolytic drugs: From discovery to translation. *Journal of Internal Medicine, 288*(5), 518–536. https://doi.org/10.1111/joim.13141

Morgunova, G. V., & Khokhlov, A. N. (2024). Drugs with senolytic activity: Prospects and possible limitations. *Moscow University Biological Sciences Bulletin, 78*(4), 268–273. https://doi.org/10.3103/S0096392524600455

Palmer, A. K., Xu, M., Zhu, Y., Pirtskhalava, T., Weivoda, M. M., Hachfeld, C. M., Prata, L. G., Casaclang-Verzosa, G., Ogrodnik, M., Schafer, M. J., White, T. A., Hickson, L. J., Tchkonia, T., Kirkland, J. L., & Passos, J. F. (2019). Targeting senescent cells alleviates obesity-induced metabolic dysfunction. *Aging Cell, 18*(3), e12950. https://doi.org/10.1111/acel.12950

Smer-Barreto, V., Quintanilla, A., Elliott, R. J., Dawson, J. C., Sun, J., Campa, V. M., Lorente-Macías, Á., Unciti-Broceta, A., Carragher, N. O., Acosta, J. C., & Oyarzún, D. A. (2023). Discovery of senolytics using machine learning. *Nature Communications, 14*, 3375. https://doi.org/10.1038/s41467-023-19045-6

Wilkinson, J. E., Burmeister, L., Brooks, S. V., Chan, C.-C., Friedline, S., Harrison, D. E., Hejtmancik, J. F., Nadon, N. L., & Miller, R. A. (2020). Acarbose has sex-dependent and -independent effects on age-related physical function, pathologies, and gut microbiota in mice. *JCI Insight, 5*(19), e137474. https://doi.org/10.1172/jci.insight.137474

Wu, B., Yan, J., Yang, J., Xia, Y., Li, D., Zhang, F., & Cao, H. (2022). Extension of the lifespan by acarbose: Is it mediated by the gut microbiota? *Aging and Disease, 13*(4), 1005–1014. https://doi.org/10.14336/AD.2022.0117

Xu, M., Palmer, A. K., Ding, H., Weivoda, M. M., Pirtskhalava, T., White, T. A., Sepe, A., Johnson, K. O., Stout, M. B., Giorgadze, N., Jensen, M. D., LeBrasseur, N. K., Tchkonia, T., & Kirkland, J. L. (2018). Targeting senescent cells enhances adipogenesis and metabolic function in old age. *eLife, 7*, e32004. https://doi.org/10.7554/eLife.32004

Senolytics: News Articles and Online Resources

Miller, R. A. (2024, December 4). *Longevity drugs, aging biomarkers, and updated findings from the Interventions Testing Program.* Peter Attia MD. https://peterattiamd.com/richardmiller2/

Ramakrishnan, V. (2024, May 15). *Aging might not be inevitable*. Wired. https://www.wired.com/story/aging-might-not-be-inevitable-wired-health-venki-ramakrishnan/

Sheloukhova, L. (2024, October 5). *Combination of rapamycin and acarbose extends lifespan*. Lifespan.io. https://www.lifespan.io/news/combination-of-rapamycin-and-acarbose-extends-lifespan/

Sheloukhova, L. (2022, October 5). *Combination of rapamycin and senolytics extends lifespan*. Lifespan.io. https://www.lifespan.io/news/combination-of-rapamycin-and-senolytics-extends-lifespan/

Harvard Gazette. (2019, November 13). *Combination gene therapy treats age-related diseases*. https://news.harvard.edu/gazette/story/2019/11/researchers-able-to-improve-reverse-age-related-diseases-in-mice

Biohacking

Anti Aging Bed. (2024). *Biohacking for longevity: Strategies to extend your lifespan and vitality*. https://antiagingbed.com/blogs/biohacking/biohacking-for-longevity-strategies-to-extend-your-lifespan-and-vitality

Business Insider. (2024). *Stem cell injections for joint rejuvenation: Exploring regenerative therapies in the longevity industry*. https://www.businessinsider.com/stem-cell-injections-knee-joints-longevity-wellness-industry-bryan-johnson-2024-7

Elliott, R. A. (2023). Bio-hacking better health: Leveraging metabolic biochemistry to enhance healthspan. *Antioxidants, 12*(9), 1749. https://doi.org/10.3390/antiox12091749

Forbes Health. (2024). *Biohacking: What is it and how does it work?* https://www.forbes.com/health/wellness/biohacking/

HCN Health. (2024). *The current state and future of biohacking*. https://hcn.health/hcn-trends-story/the-current-state-and-future-of-biohacking/

Nava Center. (2024). *Biohacking explained: Unlock a healthier, longer life*. https://navacenter.com/biohacking-explained-unlock-a-healthier-longer-life

Outliyr. (2024). *25+ exciting future biohacking trends for 2025 and beyond*. https://outliyr.com/future-biohacking-trends

Preventive Medicine Daily. (2024). *Biohacking for longevity: Enhancing healthspan through targeted lifestyle changes*. https://www.preventivemedicinedaily.com/healthy-living/biohacking/biohacking-for-longevity

Xue, C., & Li, S. (2018). Hacking aging: A strategy to use big data from medical studies to extend human lifespan. *Frontiers in Genetics, 9*, 483. https://doi.org/10.3389/fgene.2018.00483

Regeneration Therapies

Abbas, O., & Amini-Nik, S. (2025). Advances in regenerative medicine-based approaches for skin repair and rejuvenation. *Frontiers in Bioengineering and Biotechnology*. https://doi.org/10.3389/fbioe.2025.1527854

Ashammakhi, N., Reis, R. L., Chiellini, F., & Khademhosseini, A. (2024). Fundamental and practical perspectives in regenerative medicine. *International Journal of Molecular Sciences, 25*(21), 11508. https://doi.org/10.3390/ijms252111508

Atala, A., & Murphy, S. V. (2014). 3D bioprinting of tissues and organs. *Nature Biotechnology, 32*(8), 773–785. https://doi.org/10.1038/nbt.2958

Dimmeler, S., Ding, S., Rando, T. A., & Trounson, A. (2014). Translational strategies and challenges in regenerative medicine. *Nature Medicine, 20*(8), 814–821. https://doi.org/10.1038/nm.3627

Gonçalves, N. J., Gomes, C. A., Marques, A. P., & Pirraco, R. P. (2024). Achievements and future challenges for regenerative medicine. *Regenerative Therapy, 25*, 2–8. https://doi.org/10.1016/j.reth.2024.01.002

Graham, S. E., & Foutz, A. S. (2020). Stem cell therapy: Principles, advances, and challenges. *Regenerative Medicine, 15*(3), 301–317. https://doi.org/10.2217/rme-2020-0027

Loomba, R., Adams, L. A., & Schuppan, D. (2021). Gene editing and regenerative therapies for metabolic liver diseases. *Nature Reviews Gastroenterology & Hepatology, 18*(7), 463–475. https://doi.org/10.1038/s41575-021-00441-2

Patel, A., Bhartiya, D., Sharma, D., & Tiwari, V. (2024). Therapeutic approaches of cell therapy based on stem cells: A review. *Stem Cell Research & Therapy, 15*, Article 7. https://doi.org/10.1186/s13287-024-03629-0

Reardon, S. (2022). First pig-to-human heart transplant: What can scientists learn? *Nature, 601*(7894), 305–306. https://doi.org/10.1038/d41586-022-00111-9

Soto-Gutierrez, A., Wertheim, J. A., Ott, H. C., & Gilbert, T. W. (2019). Bioengineering and organ transplantation: Current status and future perspectives. *Transplantation, 103*(9), 1811–1820. https://doi.org/10.1097/TP.0000000000002816

Uddin, M. H., Dutta, S., Rahman, M. A., & Islam, M. S. (2023). Enhancing regenerative medicine: The crucial role of stem cell therapy. *Stem Cells International, 2023*, Article 5554608. https://doi.org/10.1155/2023/5554608

Xu, X., Zhang, Z., & Niu, J. (2021). Advances in senolytics: Emerging applications in aging and age-related diseases. *Nature Aging, 1*(7), 589–601. https://doi.org/10.1038/s43587-021-00087-0

Yi, X., Guo, J., Chen, M., & Shi, Y. (2024). Evolution of biotechnological advances and regenerative therapies. *Human Reproduction Update, 30*(5), 584–600. https://doi.org/10.1093/humupd/dmad005

Zhang, Y., & Zhang, Y. (2018). Tissue engineering: Principles and practice. *Annual Review of Biomedical Engineering, 20*(1), 213–236. https://doi.org/10.1146/annurev-bioeng-062117-120936

News and Institutional Reports

Alliance for Regenerative Medicine. (2023). *Cell and gene therapies poised to disrupt healthcare status quo with wave of new treatments*. https://alliancerm.org/press-release/cell-and-gene-therapies-poised-to-disrupt-health-care-status-quo-with-wave-of-new-treatments

Mayo Clinic. (2021). *Stem cell therapy: What you need to know*. https://www.mayoclinic.org/tests-procedures/stem-cell-therapy/about/pac-20384540

News and Market Insights

iPSC News. (2024). *20 stem cell and regenerative medicine predictions for 2024*. https://ipscell.com/2024/01/20-stem-cell-regenerative-medicine-predictions-for-2024

iPSC News. (2024). *25 stem cell and regenerative medicine predictions for 2025*. https://ipscell.com/2024/12/25-stem-cell-regenerative-medicine-predictions-for-2025

MarketWatch. (2024). *RFK Jr. could prove a surprise boon for stem-cell stocks with pivotal year ahead*. https://www.marketwatch.com/story/rfk-jr-could-prove-a-surprise-boon-for-stem-cell-stocks-with-pivotal-year-ahead-37247538

The Times. (2024). *The "holy grail" of heart health: A valve that grows inside you*. https://www.thetimes.com/uk/healthcare/article/the-holy-grail-of-heart-health-a-valve-that-grows-inside-you-pjxvrq331

The Wall Street Journal. (2024). *Science is finding ways to regenerate your heart*. https://www.wsj.com/health/grow-heart-lung-tissue-medical-technology-24b22bb4

Gene Therapy and Aging

Davidsohn, N., Pezone, M., Vernet, A., Graveline, A., Oliver, D., Slomovic, S., & Mooney, D. J. (2019). A single combination gene therapy treats multiple age-related diseases. *Proceedings of the National Academy of Sciences, 116*(42), 20817–20819. https://doi.org/10.1073/pnas.1910073116

Doudna, J. A., & Charpentier, E. (2014). The new frontier of genome engineering with CRISPR-Cas9. *Science, 346*(6213), 1258096. https://doi.org/10.1126/science.1258096

Harvard Gazette. (2019, November 13). *Combination gene therapy treats age-related diseases.* https://news.harvard.edu/gazette/story/2019/11/researchers-able-to-improve-reverse-age-related-diseases-in-mice

Kenyon, C. J. (2010). The genetics of ageing. *Nature, 464*(7288), 504–512. https://doi.org/10.1038/nature08980

Kim, J., Moon, S. Y., Kang, H. G., Kim, H. J., Choi, J. S., Lee, S. H. S., Park, K., & Won, S.-Y. (2025). Therapeutic potential of AAV2-shmTOR gene therapy in reducing retinal inflammation and preserving endothelial integrity in age-related macular degeneration. *Scientific Reports, 15*(1), Article 9517. https://doi.org/10.1038/s41598-025-93993-4

Koh, D., Kim, Y., Park, J., Lee, S., & Choi, E. (2025). Reduced UPF1 levels in senescence impair nonsense-mediated mRNA decay. *Communications Biology, 8*, Article 123. https://doi.org/10.1038/s42003-025-04567-8

Kumar, S., & Gupta, R. (2025). Chemical enhancement of DNA repair in aging. *bioRxiv.* https://doi.org/10.1101/2025.02.21.639496

Leung, M. K., Delong, A., Alipanahi, B., & Frey, B. J. (2016). Machine learning in genomic medicine: A review of computational problems and data sets. *Proceedings of the IEEE, 104*(1), 176–197. https://doi.org/10.1109/JPROC.2015.2494198

New York Post. (2025, January 4). *Researchers discover aging "hotspot" in the brain—and it could have big implications for patients.* https://nypost.com/2025/01/04/health/researchers-discover-aging-hotspot-in-the-brain-and-the-fine-could-have-big-implications-for-patients

Ocampo, A., Reddy, P., Martinez-Redondo, P., Platero-Luengo, A., Hatanaka, F., Hishida, T., … Izpisua Belmonte, J. C. (2016). In vivo amelioration of age-associated hallmarks by partial reprogramming. *Cell, 167*(7), 1719–1733.e12. https://doi.org/10.1016/j.cell.2016.11.052

Ruffo, P., Traynor, B. J., & Conforti, F. L. (2025). Advancements in genetic research and RNA therapy strategies for amyotrophic lateral sclerosis (ALS): Current progress and future prospects. *Journal of Neurology, 272*(3), 233. https://doi.org/10.1007/s00415-025-12975-8

SciTechDaily. (2023, September 1). *Longevity breakthrough: New treatment reverses multiple hallmarks of aging.* https://scitechdaily.com/longevity-breakthrough-new-treatment-reverses-multiple-hallmarks-of-aging

Sun, N., Youle, R. J., & Finkel, T. (2016). The mitochondrial basis of aging. *Molecular Cell, 61*(5), 654–666. https://doi.org/10.1016/j.molcel.2016.01.028

Tharmapalan, V., Jones, M., & Smith, L. (2025). Senolytic compounds reduce epigenetic age of blood samples in vitro. *npj Aging, 1*, Article 45. https://doi.org/10.1038/s41514-025-00321-9

Wired. (2025, January 7). *Correcting genetic spelling errors with next-generation CRISPR.* https://www.wired.com/story/correcting-genetic-spelling-errors-with-next-generation-crispr

Technology and Aging

Aleppo, G., & Ruedy, K. J. (2017). Role of continuous glucose monitoring in diabetes treatment. *Diabetes Technology & Therapeutics, 19*(S2), S9–S14.

Antony, V. N., Jeon, C., Li, J., Gao, G., Peng, H., Ostrowski, A. K., & Huang, C.-M. (2025). The design of on-body robots for older adults. *arXiv preprint.* https://arxiv.org/abs/2502.02725

Bumgarner, J. M., et al. (2020). Smartwatch algorithm for automated detection of atrial fibrillation. *Journal of the American College of Cardiology, 75*(10), 1145–1154. https://doi.org/10.1016/j.jacc.2019.12.054

de Zambotti, M., et al. (2019). Wearable sensors for sleep monitoring. *Sleep Medicine Clinics, 14*(1), 75–92. https://doi.org/10.1016/j.jsmc.2018.10.003

Doudna, J. A., & Sternberg, S. H. (2017). *A crack in creation: Gene editing and the unthinkable power to control evolution.* Houghton Mifflin Harcourt.

Esteva, A., Kuprel, B., Novoa, R. A., Ko, J., Swetter, S. M., Blau, H. M., & Thrun, S. (2017). Dermatologist-level classification of skin cancer with deep neural networks. *Nature, 542*(7639), 115–118. https://doi.org/10.1038/nature21056

Famm, K., Litt, B., Tracey, K. J., Boyden, E. S., & Slaoui, M. (2013). Drug discovery: A jump-start for electroceuticals. *Nature, 496*(7444), 159–161. https://doi.org/10.1038/496159a

Ferguson, T., Rowlands, A. V., Olds, T., & Maher, C. (2015). Validity of consumer-level activity monitors in free-living conditions. *International Journal of Behavioral Nutrition and Physical Activity, 12*, Article 135. https://doi.org/10.1186/s12966-015-0314-4

Gajarawala, S. N., & Pelkowski, J. N. (2021). Telehealth benefits and barriers. *Journal for Nurse Practitioners, 17*(2), 218–221. https://doi.org/10.1016/j.nurpra.2020.09.013

Granato, D., et al. (2020). Functional foods: Current status and future trends. *Annual Review of Food Science and Technology, 11*, 93–118. https://doi.org/10.1146/annurev-food-032519-051708

Heron, K. E., & Smyth, J. M. (2010). Ecological momentary interventions and mobile technology. *British Journal of Health Psychology, 15*(1), 1–39. https://doi.org/10.1348/135910709X466063

Kirkland, J. L., Tchkonia, T., Zhu, Y., Niedernhofer, L. J., & Robbins, P. D.

(2017). The clinical potential of senolytic drugs. *Journal of the American Geriatrics Society, 65*(10), 2297–2301. https://doi.org/10.1111/jgs.14969

Lee, T., & Zhang, Y. (2024). Genomic platforms and longevity: Identifying pro-longevity markers. *arXiv preprint*. https://arxiv.org/abs/2403.19087

Li, A., Montaño, Z., Chen, V. J., & Gold, J. I. (2011). Virtual reality and pain management. *Pain Management, 1*(2), 147–157. https://doi.org/10.2217/pmt.10.15

Murphy, S. V., & Atala, A. (2014). 3D bioprinting of tissues and organs. *Nature Biotechnology, 32*(8), 773–785. https://doi.org/10.1038/nbt.2958

Relling, M. V., & Evans, W. E. (2015). Pharmacogenomics in the clinic. *Nature, 526*(7573), 343–350. https://doi.org/10.1038/nature15817

Rizzo, A., & Koenig, S. (2017). Is clinical virtual reality ready for primetime? *Neuropsychology, 31*(8), 877–899. https://doi.org/10.1037/neu0000405

Sun, Y., Ankenbauer, S. A., Guo, Z., Chen, Y., Ma, X., & He, L. (2025). Rethinking technological solutions for community-based older adult care. *arXiv preprint*. https://arxiv.org/abs/2503.23609

Topol, E. J. (2019). High-performance medicine: The convergence of human and artificial intelligence. *Nature Medicine, 25*(1), 44–56. https://doi.org/10.1038/s41591-018-0300-7

Tomašev, N., et al. (2019). Continuous prediction of acute kidney injury. *Nature, 572*(7767), 116–119. https://doi.org/10.1038/s41586-019-1390-1

Trounson, A., & McDonald, C. (2015). Stem cell therapies in clinical trials. *Cell Stem Cell, 17*(1), 11–22. https://doi.org/10.1016/j.stem.2015.06.007

Institutional and News Sources

Forbes Technology Council. (2024, June 28). *Unlocking "healthspan" with technology can help people live longer and healthier lives*. https://www.forbes.com

identifyHer. (2025, January 7). *Peri: A wearable device for perimenopause symptom tracking*. The Verge. https://www.theverge.com/2025/1/7/24337603/identifyher-peri-ces-2025-perimenopause-wearable-health-tech

Yang, J., Petty, C. A., Dixon-McDougall, T., Lopez, M. V., Tyshkovskiy, A., Maybury-Lewis, S., Tian, X., Ibrahim, N., Chen, Z., Griffin, P. T., Arnold, M., Li, J., Martinez, O. A., et al. (2023). Chemically induced reprogramming to reverse cellular aging. *Aging (Albany NY), 15*, 5966–5989. https://doi.org/10.18632/aging.204896

Zhu, Y., et al. (2015). The Achilles' heel of senescent cells: From transcriptome to senolytic drugs. *Aging Cell, 14*(4), 644–658. https://doi.org/10.1111/acel.12344

Artificial Intelligence

Aliper, A., & Zhavoronkov, A. (2018). Deep learning applications for predicting pharmacological properties of drugs and drug repurposing using transcriptomic data. *Molecular Pharmaceutics, 15*(12), 4311–4320. https://doi.org/10.1021/acs.molpharmaceut.8b00839

Bischof, E., & Wegrzyn, D. (2024). The Longevity Med Summit: Insights on healthspan from cell to society. *Frontiers in Aging, 5*, Article 1417455. https://doi.org/10.3389/fragi.2024.1417455

Cleveland Clinic. (2024). *AI in healthcare: Enhancing diagnosis and treatment.* https://health.clevelandclinic.org

Ferrucci, L., & Gonzalez-Freire, M. (2021). The use of biomarkers and artificial intelligence in the study of aging and longevity. *Aging Clinical and Experimental Research, 33*(6), 1407–1413. https://doi.org/10.1007/s40520-021-01868-w

Insilico Medicine. (2024). Advancing AI in drug discovery for age-related diseases: IPO filing and Phase IIa results. *Lifespan.io.* https://www.lifespan.io/news/ai-in-longevity-the-reality-today

Kaeberlein, M., & Kennedy, B. K. (2023). Artificial intelligence and the future of longevity science. *Science Translational Medicine, 15*(670), eabc8732. https://doi.org/10.1126/scitranslmed.abc8732

Kosheleva, N., Shadrina, M. S., & Zhavoronkov, A. (2021). Predicting lifespan and healthspan from physiological and biological data using artificial intelligence. *Current Opinion in Systems Biology, 27*, 100355. https://doi.org/10.1016/j.coisb.2021.07.001

Lifespan.io. (2024). *What AI technology is doing for longevity now.* https://www.lifespan.io/news/what-ai-technology-is-doing-for-longevity-now

Mamoshina, P., Kochetov, K., Putin, E., & Cortese, F. (2023). Towards AI-driven longevity research: An overview. *Frontiers in Aging, 4*, Article 1057204. https://doi.org/10.3389/fragi.2023.1057204

Mayo Clinic Press. (2024). *AI in healthcare: The future of patient care and health management.* https://mcpress.mayoclinic.org

MDPI. (2024). Accelerating drug discovery with AI: Reducing time and cost. *Pharmaceutics, 16*(10), 1328. https://doi.org/10.3390/pharmaceutics16101328

MedWave.io. (2024). *How AI is transforming healthcare: 12 real-world use cases.* https://medwave.io/2024/01/how-ai-is-transforming-healthcare-12-real-world-use-cases

Shaban-Nejad, A., Michalowski, M., & Bianco, S. (2023). Artificial intelligence for personalized care, wellness, and longevity research. In A. Shaban-Nejad,

M. Michalowski, & S. Bianco (Eds.), *Artificial intelligence for personalized medicine* (Studies in Computational Intelligence, Vol. 1106). Springer, Cham. https://doi.org/10.1007/978-3-031-36938-4_1

Silcox, C., Zimlichmann, E., Huber, K., Rowen, N., Saunders, R., McClellan, M., Kahn, C. N., Salzberg, C. A., & Bates, D. W. (2024). The potential for artificial intelligence to transform healthcare: Perspectives from international health leaders. *npj Digital Medicine*, 7, Article 88. https://doi.org/10.1038/s41746-024-00888-7

Singh, N., & Peden, J. (2023). Artificial intelligence for personalized aging and longevity interventions: Current trends and future perspectives. *Frontiers in Aging Neuroscience*, 15, 998233. https://doi.org/10.3389/fnagi.2023.998233

Stanford Institute for Human-Centered Artificial Intelligence. (2024). *AI Index Report 2024*. Stanford University. https://aiindex.stanford.edu/report/

Time. (2024). *How AI behavior change applications are improving health care*. https://time.com/6994739/ai-behavior-change-health-care

World Economic Forum. (2024). *AI in healthcare: Diagnostics and improved health outcomes*. https://www.weforum.org/stories/2024/09/ai-diagnostics-health-outcomes

Zhavoronkov, A., Aliper, A., & Artal-Sanz, M. (2023). Aging clocks and xAI: Enhancing aging research with interpretable artificial intelligence. *Nature Aging*, 3(2), 150–159. https://doi.org/10.1038/s43587-023-00233-9

Zhavoronkov, A., Bischof, E., & Lee, K.-F. (2021). Artificial intelligence in longevity medicine. *Nature Aging*, 1(1), 5–7. https://doi.org/10.1038/s43587-020-00011-8

Zhavoronkov, A., Mamoshina, P., Vanhaelen, Q., Scheibye-Knudsen, M., Moskalev, A., & Aliper, A. (2019). Artificial intelligence for aging and longevity research: Recent advances and perspectives. *Ageing Research Reviews*, 49, 49–66. https://doi.org/10.1016/j.arr.2018.11.003

Partnering with Your Healthcare Provider

Being Brigid. (2024). *Health optimization through personalized care and proactive strategies*. https://beingbrigid.com/health-optimization

Lifespan.io. (2024). *Insights from the roundtable of longevity clinics 2024: Advancing healthspan through personalized healthcare*. https://www.lifespan.io/news/insights-from-the-roundtable-of-longevity-clinics-2024

Longevity Technology. (2024). *What the future of longevity looks like with Lifeforce: Partnering with providers for better health outcomes*. https://longevity.technology/news/what-the-future-of-longevity-looks-like-with-lifeforce

Sanatorium Health. (2024). *Unlocking longevity: Essential doctor *visits for optimal health*. https://www.sanatorium.health/unlocking-longevity-essential-doctor-visits-for-optimal-health

Time. (2024). *Annual physical: What to expect during a doctor's appointment*. https://time.com/7098720/annual-physical-what-to-expect-doctors-appointment

Time. (2024). *Healthspan vs. lifespan: What's the difference? The role of healthcare in living longer and healthier lives*. https://time.com/6341027/what-is-healthspan-vs-lifespan

Two Eleven Health. (2024). *Building a doctor-patient relationship: Key to optimal health outcomes*. https://twoelevenhealth.com/blog/building-a-doctor-patient-relationship

WebMD. (2024). *What is holistic medicine? A guide to whole-person care*. https://www.webmd.com/balance/what-is-holistic-medicine

Creating a Longevity Plan

BalanceGenics. (2023). *The 9 longevity secrets from the world's longest-lived people, according to Dan Buettner*. https://balancegenics.com/blogs/post/the-9-longevity-secrets-from-the-world-s-longest-lived-people-according-to-dan-buettner

BetterUp. (2023). *Life planning: The ultimate guide (and template) to achieving your goals*. https://www.betterup.com/blog/life-planning

Cona, L. A. (2024). *How to live longer: A guide to longevity*. DVC Stem. https://www.dvcstem.com/post/how-to-live-longer

Harvard Health Publishing. (2023). *Longevity: Lifestyle strategies for living a healthy, long life*. https://www.health.harvard.edu/staying-healthy/longevity-lifestyle-strategies-for-living-a-healthy-long-life

Lifehack. (2023). *How to create a life plan (with action plan and tips)*. https://www.lifehack.org/886768/how-to-make-a-life-plan

Longevity Training Academy. (2024). *Top ten longevity strategies for 2024*. https://longevitytrainingacademy.com/top-ten-longevity-strategies-for-2024/

MindGrow. (2023). *The ultimate guide to longevity: 5 key practices for a longer, healthier life*. https://mindgrow.io/the-ultimate-guide-to-longevity-5-key-practices-for-a-longer-healthier-life

PositivePsychology.com. (2023). *How to create a personal development plan: 3 examples*. https://positivepsychology.com/personal-development-plan/

Sanatorium Health. (2024). *Top 10 longevity leaders reveal new secrets*. https://www.sanatorium.health/sep-2024-top-10-longevity-leaders-reveal-new-secrets/

Stanford Center on Longevity. (2017). *Life planning in the age of longevity: Action plan*. https://longevity.stanford.edu/wp-content/uploads/dlm_uploads/2017/05/Life-Planning-Action-Plan.pdf

"Begin at th*e beginning and go on till you come to the end; then stop.*"
—Lewis Carroll, Alice in Wonderland

www.ingramcontent.com/pod-product-compliance
Lightning Source LLC
Chambersburg PA
CBHW020533030426
42337CB00013B/831